Lecture Notes
in Control and Information Sciences 257

Editor: M. Thoma

Springer-Verlag London Ltd.

M. Moallem, R.V. Patel and K. Khorasani

Flexible-link Robot Manipulators

Control Techniques and Structural Design

With 40 Figures

Springer

Authors

M. Moallem, PhD
R.V. Patel, PhD
University of Western Ontario, London, Ontario, Canada

K.Khorasani, PhD
Department of Electrical Engineering, Concordia University, Montreal, PQ,
Canada

ISBN 978-1-85233-333-1

British Library Cataloguing in Publication Data

Moallem, M.
    Flexible-link robot manipulators : control technique and
    Structural design. - (Lecture notes in control and
    information sciences ; 257)
    1.Robots 2.Manipulators (Mechanism)
    I.Title II.Patel, Rajnikant V. III.Khorasani, K.
    629.8'92
    ISBN 978-1-85233-333-1   ISBN 978-1-84628-555-4 (eBook)
    DOI 10.1007/978-1-84628-555-4
Library of Congress Cataloging-in-Publication Data
A catalog record for this book is available from the Library of Congress

Typesetting: Camera ready by authors

69/3830-543210 Printed on acid-free paper SPIN 10768749

Preface

This monograph is concerned with development and implementation of nonlinear mathematical techniques for feedback control and shape design of robot manipulators whose links have considerable structural flexibility. A flexible-link manipulator is an underactuated dynamic system in the sense that it has more degrees of freedom than its actuation inputs. In a *fully-actuated* system such as a rigid-link robot manipulator, for *each* actuation input, i.e., servomotor, there is only *one* degree of freedom. This condition is not satisfied for underactuated systems. For example, due to the prohibitive cost of placing equipment in outer space, space-based robot manipulators are comprised of long, lightweight, flexible links. Compared to the *rigid* case, the servomotors should in this case provide torques to handle deflections and/or vibrations of the links in addition to providing inputs for moving the joints. Serious problems arise in controlling underactuated systems compared to fully actuated systems.

Underactuated mechanical systems with structural flexibility are not restricted to space-based robotic applications. There are interesting applications of these systems such as macro-micro manipulators in aerospace, electrical power transmission line maintenance, nuclear waste disposal, support systems in imaging technologies, microelectronic wafer fabrication, space antennas, solar panels, acoustical systems, underwater applications, etc, to name a few. All of these systems, more or less, manifest similar behavior since they obey similar natural laws of Lagrangian systems. In general, the above systems exhibit characteristics such as nonlinearities, unstable internal dynamics, parametric and dynamic uncertainties, stiff dynamic behavior, and other properties that place them in the forefront of research in electromechanical systems. This class of mechanical systems is rich in both applications and control problems and, unlike fully actuated systems, there are generally few results that are applicable to all of them.

The study of dynamic properties and control techniques for flexible-link manipulators can also be a framework for designing the mechanical shape and material of these systems such that improved properties can be achieved in order to facilitate the control problem. This aspect of design has been followed in other areas such as designing the shape of an aircraft for achieving stable open-loop behavior, but is still not widely developed and investigated in flexible-link manipulators.

The organization of this monograph is as follows. In Chapter 1, the literature review covers different methods that have been proposed for control and design of flexible-link manipulators. This chapter also introduces some basic concepts and definitions such as *internal dynamics*, *zero dynamics*, and the *nonminimum-phase characteristic* for nonlinear systems. Chapter 2 is concerned with two-time scale modeling and control of flexible-link manipulators and is based on the concept of integral manifolds in singular perturbation theory. Chapter 3 discusses the nonlinear input-output decoupling control. The practical implementation issues for this control technique require the inconvenient measurement of flexural rates. Towards this end, an observation strategy is proposed in Chapter 4 to circumvent this problem. Experimental evaluation of the control strategies in Chapters 2 to 4 are carried out on a setup constructed in the laboratory and implemented under the Chimera 3.1 real-time operating system. In Chapter 5, we consider the use of sliding control as a means of dealing with modeling uncertainties in the dynamic model of the system. In Chapter 6, structural shape optimization is considered as a means of improving the dynamic behavior. In this regard, a multi-objective optimization procedure is introduced for designing the arm shape such that some desired characteristics are achieved based on the controllability and observability of flexible modes. Finally, stability proofs for the control schemes presented in this monograph are presented in the appendices.

Acknowledgements

The authors would like to acknowledge the numerous researchers who have, directly or indirectly, contributed to the field of flexible-link manipulators. The authors have made every attempt to include references to related work and to give credit to other researchers where appropriate. Any omission is entirely unintentional and the authors would be pleased to receive information about or copies of relevant publications that have not been referenced in this monograph.

The authors are indebted to the Natural Sciences and Engineering Research Council (NSERC) of Canada, the Fonds pour la Formation de Chercheurs et l'Aide à la Recherche (FCAR) of the Province of Quebec, and the University of Western Ontario, for funding much of this research, through research grants and startup funds.

Finally, the authors would like to thank their families for their support and understanding throughout this project.

M. Moallem
R.V. Patel
K. Khorasani

Contents

Preface   **V**

1. Introduction   **1**
  1.1. Motivation   2
  1.2. Dynamic Modeling   4
  1.3. Control Strategies 10
  1.4. Mechanical Design 14
  1.5. Some Related Topics 16
    1.5.1. Regulation and Tracking 16
    1.5.2. Some Examples of Non–minimum Phase Systems 20
  1.6. Outline of the Monograph 26
    1.6.1. Tracking Control Using Integral Manifolds . . . 27
    1.6.2. Tracking Control by Output Redefinition and
         Input–output Decoupling 28
    1.6.3. Observation Strategy for Flexural Rates 28
    1.6.4. Structure Design 28
    1.6.5. Experimental Evaluation 29

2. Tracking Control by Integral Manifolds   **31**
  2.1. Introduction 32
  2.2. Model Reduction Using Integral Manifolds 33

2.3. Slow and Fast Subsystem Control Strategies 38
2.4. Experimental Results 48
    2.4.1. Model Validation 50
    2.4.2. Implementation of the Control Law 52
2.5. Conclusion 56

3. Decoupling Control   **57**
3.1. Introduction 57
3.2. Input–output Linearization 58
    3.2.1. Derivation of H(0,0) 66
    3.2.2. A Model for the Damping Term E_2 67
3.3. Case Studies 67
    3.3.1. Regions having the Minimum–phase Property
        for a Two–link Manipulator 67
    3.3.2. Inverse–dynamics Control 69
3.4. Conclusion 71

4. Observer–based Decoupling Control   **73**
4.1. Introduction 73
4.2. Inverse–dynamics Control 75
4.3. Observer Design 76
    4.3.1. Full–order Observer 76
    4.3.2. Reduced–order Observer 78
    4.3.3. Sliding Observer 79
4.4. Observer Based Inverse–dynamics Control 80
4.5. Implementation of the Control Law 82
    4.5.1. Experimental Results 86
4.6. Conclusion 89

5. Inverse Dynamics Sliding Control   **95**
5.1. Introduction 95
5.2. Control Based on Input–output Linearization 96
    5.2.1. Stability of the Closed–loop System 99
    5.2.2. Numerical Simulations 100
5.3. Conclusion 102

6. Optimum Structure Design for Control   **103**
6.1. Introduction 104
6.2. Flexural Dynamics and Eigenvalue Sensitivities 105
    6.2.1. Controllability and Sensitivity of Flexural Modes 107

   6.2.2. Pole–zero Relationships 109
  6.3. Multi–objective Optimization 112
  6.4. Optimization Procedure and Discussion of Results . . 113
   6.4.1. Performance Comparison between Non–uniform
     and Uniform Arms 115
  6.5. Conclusion 119

7. Concluding Remarks   **121**
  7.1. Control Using Integral Manifolds 122
  7.2. Input–output Decoupling Control 122
  7.3. Structure Design 122

A. Stability Proofs   **125**
  A.1. Proof of Theorem 2.1 125
  A.2. Proof of Theorem 3.1 128
  A.3. Proof of Theorem 4.1 131
  A.4. Proof of Theorem 5.1 134

B. Kinematic Description   **137**

C. Dynamic Models   **139**
  C.1. Dynamic Model of the Flexible Single–link Arm 139
  C.2. Dynamic Model of the Flexible Two–link Arm 140
  C.3. Derivation of the Dynamic Equations 141

References   **147**

Index   **157**

1.

Introduction

A flexible-link robot manipulator consists of a series of structurally flexible arms connected by joints to form a spatial mechanism. In most cases, structural flexibility in not an intended design feature of the manipulator system. Flexibility can be a consequence of constructing the manipulator from light materials which itself can result from constraints on inertia of the mechanism or manipulation speed. However, structural flexibility may be intentionally introduced in manipulator design to make the robot system more compliant with its environment. Even robot manipulators considered as *rigid* have inherent structural flexibilities which can cause stability problems when carrying large payloads with high speeds.

The flexural effects combined with the inherent nonlinear dynamics of flexible-link manipulators make the control problem of such systems more difficult. Therefore, the control and design of flexible-link manipulators remains an open and challenging problem in robotics research. From the control point of view, one has to deal with the nonlinear, coupled, and stiff dynamics of the system. A challenging problem in this regard is to cause the output of the system to track a desired trajectory while maintaining internal stability of a nominally unstable system. From a design perspective, one is concerned with designing the robot geometric structure and material such that while certain mechanical and dynamic criteria are met, the control problem is also simplified.

This chapter is an introduction to issues related to modeling, control, and design of flexible–link robotic manipulators. The philosophy and technical difficulties associated with these robots are illustrated in Section 1.1.   Section 1.2 covers a literature review on the modeling of flexible–link manipulators. By modeling, we mean the plant dynamics that are used for the purpose of control. The control strategies are outlined in Section 1.3 and the design problem is discussed in Section 1.4.   In Section 1.5, we consider some topics related to the control problem of these robots. Some examples are also given from other fields which have similar difficulties from the control point of view. Finally, the contributions of this monograph are stated briefly in Section 1.6.

1.1.   Motivation

The modeling, design, and motion control of structurally flexible robotic manipulators have been the focus of attention of researchers in the past few years. This field of research has been attractive for several reasons. For instance, space applications require low–mass designs, to achieve escape velocity, and in order to accomplish a mission with better fuel economy. This restriction puts a limitation on the degree of rigidity of space robots. On the other hand, increased structural flexibility may be desirable in tasks such as cleaning delicate surfaces, or avoiding damages to the manipulator system due to accidental collisions. Still the problem may be viewed from other angles such as high–speed manipulation and increased productivity. Conventional manipulators are limited to a load–carrying capacity of 5-10 percent of their own weight. This restriction is mainly due to the requirement of having a stable closed–loop system. This has encouraged the design of heavy (rigid) robots so as to have less stringent control problems. Thus, the designers of earth-bound robots have solved the control problem by making the manipulators more rigid using bulkier structures. Likewise, for light–weight space-based systems, one simple solution is to move the manipulators slowly.

Achieving exact, high–speed manipulation with lightweight structures, is definitely a desirable objective. Such an achievement is also attractive from energy consumption considerations since smaller actuators are needed due to lighter loads. Regardless of the reason that flexibilities become significant, precise and stable control of the manipulator tip is desirable. This requires the inclusion of deformation effects due to the flexibility of the arms in the dynamic equations, and generally tends to complicate the analysis and design of the control laws. In flexible link manipulators, a major difficulty arises when one tries to track a specific tip position trajectory by applying the torques at the actuation ends. Due to the non-colocated nature of sensors and actuators, the zero-dynamics of the system become unstable. The zero-dynamics are defined as the internal system dynamics when the outputs are driven to zero by specific inputs. The system is called non-minimum phase when these dynamics are unstable. This intrinsic non-minimum phase property hinders exact asymptotic tracking of a desired tip trajectory if causal controllers are employed. Thus in practice one may be satisfied with small (rather than zero) tracking errors. More details on this are given in Section 1.5.

This monograph aims to address two major issues regarding flexible link manipulators: Plant design and control design. The plant to be controlled is the robotic manipulator and we are interested in devising control strategies that take into account the flexural effects as well. The first aim is to alter the plant characteristics such that the final plant has some desirable features from the control point of view. This should be achieved without any sacrifice to some specifications such as the total mass or the moments of inertia experienced by the actuators. There is certainly a limit to which improvement may be achieved, and the rest of the effort is the job of the controller. Thus we are concerned with the mechanical design of the manipulator on the one hand and controller design on the other hand. By mechanical design, we mean improved structural shapes. The type of material used in design is a key issue and plays an important role. A material with a high modulus of elasticity, a low mass density, and high structural damping is the solution to most problems of flexibility. Here it is assumed that the type of material is specified. This specification may be considered as the ultimate restriction put forth either by materials technology or economic considerations. The other side of the coin is the control problem. In this respect, a dynamic model that describes the system behavior in a concise and accurate way is desirable. Improved dynamic modeling allows for reliable design and control. Theoretically, the dynamic equations are infinite dimensional and may be described by partial differential equations.

This is simply due to the fact that an infinite number of coordinates are required to kinematically describe each link. However, an infinite dimensional model may not be suitable for control system design. This is due to factors such as the dynamic model complexity and the band–limited nature of sensors and actuators. In the modeling phase, one usually truncates the number of flexible coordinates. In any case, the dynamics turn out to be a highly nonlinear and coupled set of differential equations which enjoy a two–time–scale nature. Another major question in controller design related to the number and type of sensing points. Since tip positions and their rates of change with time are to be controlled, the least information to be provided to the controller is accurate information on tip positions. In particular, tip position information can be provided by using camera vision or strain gauge measurements, however tip rate measurements are not directly obtainable.

1.2.   Dynamic Modeling

Theoretically, a flexible-link manipulator system is an infinite dimensional system since the dynamics of the system can be obtained by a set of partial differential equations. This approach to modeling can become cumbersome even for a single-link flexible arm. For the multi-link case, the boundary conditions are time-varying and a closed form solution may not be available. For this reason, discretization methods have been widely used for modeling the dynamic behavior. In practice this approach is more meaningful if we consider the band-limited nature of sensors and actuators. Knowledge of the dynamic model of a structurally flexible manipulator is needed for the purpose of simulation (forward dynamics), inverse dynamics and more importantly, for the design of the controller. For rigid–link manipulators, it is known that once the kinematics of the chain of rigid bodies are properly defined, the set of dynamic equations can be obtained by conventional methods such as Newton–Euler or Lagrangian [32]. In principle, all sets of dynamic equations that are derived based on different kinematic descriptions are equivalent. However, this is not usually true for flexible–link manipulators. The main reason is that the actual system dynamics are infinite dimensional because of the distributed link flexibility. To simplify the problem, a conventional method is to approximate the kinematics of the flexible–link system by a finite number of degrees of freedom. This can be achieved by approximating the deflection of each link by a weighted summation of modes given by

$$w(x, t) = \sum_{i=1}^{n} \phi_i(x)\delta_i(t) \qquad (1.1)$$

where x is the distance along the neutral axis of the link, $\phi_i(x)$ is the modal shape function, and δ_i is the modal variable. The modal shape functions should be differentiable functions of at least the degree of the differential equation, must satisfy the geometric boundary conditions (e.g. slope and deflection at the end points of each link), and must form a complete coordinate basis so that they can describe any link profile [15]. Additionally, if the modal shape functions form an orthogonal set, many cross terms are eliminated in the dynamic equations and simpler models are thus obtained. The dynamic equations can then be obtained by obtaining the potential (P) and kinetic (K) energies and using Lagrange's equation

$$\frac{d}{dt}\left(\frac{\partial L}{\partial \dot{x}_i}\right) - \frac{\partial L}{\partial x_i} = Q_i, \quad i = 1, \cdots, m+n \qquad (1.2)$$

where x_i is either a modal coordinate, denoted by δ_i ($i = 1, \cdots, m$), or a rigid-body coordinate, denoted by q_i ($i = 1, \cdots, n$); $L = K - P$ is the Lagrangian, and Q_i is the generalized force (see e.g. [15]). Representing $q^T = [q_1, q_2, \cdots, q_n]$, and $\delta^T = [\delta_1, \delta_2, \cdots, \delta_n]$, the dynamic equations obtained from the the above procedure have the following form

$$M(q, \delta)\begin{bmatrix} \ddot{q} \\ \ddot{\delta} \end{bmatrix} + \begin{bmatrix} f_1(q, \dot{q}) + g_1(q, \dot{q}, \delta, \dot{\delta}) + h_1(q, \delta) \\ f_2(q, \dot{q}) + g_2(q, \dot{q}, \delta, \dot{\delta}) + h_2(q, \delta) + K\delta \end{bmatrix}$$

$$= B\begin{bmatrix} u \\ 0 \end{bmatrix} \qquad (1.3)$$

where $M(q, \delta)$ is the positive–definite mass matrix, $f_1, f_2, g_1,$ and g_2 are the terms due to gravity, Coriolis, and centripetal forces, h_1 and h_2 are the terms due to gravity, K is the positive–definite stiffness matrix, u is the $n \times 1$ vector of input torques, and B is the constant input effect matrix which depends on the modal shape functions used.

Now depending on the type of modal shape functions and the number of modal variables used, the dynamics represented by the above equations can be different. For the purpose of computational efficiency in real-time control, a desirable dynamic model should express the dynamic behavior by using a small number of modes with sufficient accuracy. In what follows we review some of these approaches to modeling. It should be emphasized that we are interested in models that are suitable for control purposes. Undoubtedly, more exact modeling techniqus, such as finite elements, can be utilized to verify the accuracy of the proposed schemes.

A great deal of research in modeling and control of flexible-link robots has focussed on a single link flexible beam free to rotate and flex in a horizontal plane (see Figure 1.1). Neglecting second order effects such as rotary inertia, shear deformation and actuator dynamics, the dynamics of the position $y(x,t)$ of any point on the beam is given (e.g.[15]) by the Euler-Bernoulli beam equation

$$EI\frac{\partial^4 y(x,t)}{\partial x^4} + m\frac{\partial^2 y(x,t)}{\partial t^2} = 0 \qquad 0 < x < l \qquad (1.4)$$

where E is Young's modulus of elasticity of the material, I is the area moment of inertia of the cross section about the z–axis, m is the mass per unit length of the arm, and l is the length of the arm. Note that $y(x,t) = \theta x + w(x,t)$, and the hub angle $\psi = \frac{\partial y(0,t)}{\partial x} = \theta + \frac{\partial w}{\partial x}|_{x=0}$. The boundary conditions of (1.4) at $x = 0$ are as follows

$$y(0,t) = 0 \qquad\qquad (1.5)$$

$$EI\frac{\partial^2 y}{\partial x^2} + u = J_h\frac{d^2}{dt^2}\left(\frac{\partial y}{\partial x}\right) \qquad (1.6)$$

where J_h is the hub inertia and u is the actuating torque, and at $x = l$

$$Bending\ Moment:\ EI\frac{\partial^2 y}{\partial x^2} = -Jp\frac{d^2}{dt^2}\left(\frac{\partial y}{\partial x}\right) \qquad (1.7)$$

$$Shear\ Force:\ EI\frac{\partial^3 y}{\partial x^3} = M_p\frac{d^2}{dt^2}y \qquad (1.8)$$

where J_p, M_p are the payload inertia and mass respectively.

The flexible–link depicted in Figure (1.1) has been the subject of many research projects. Cannon and Schmitz [2] used the assumed modes method to model the flexible link. They have assumed zero

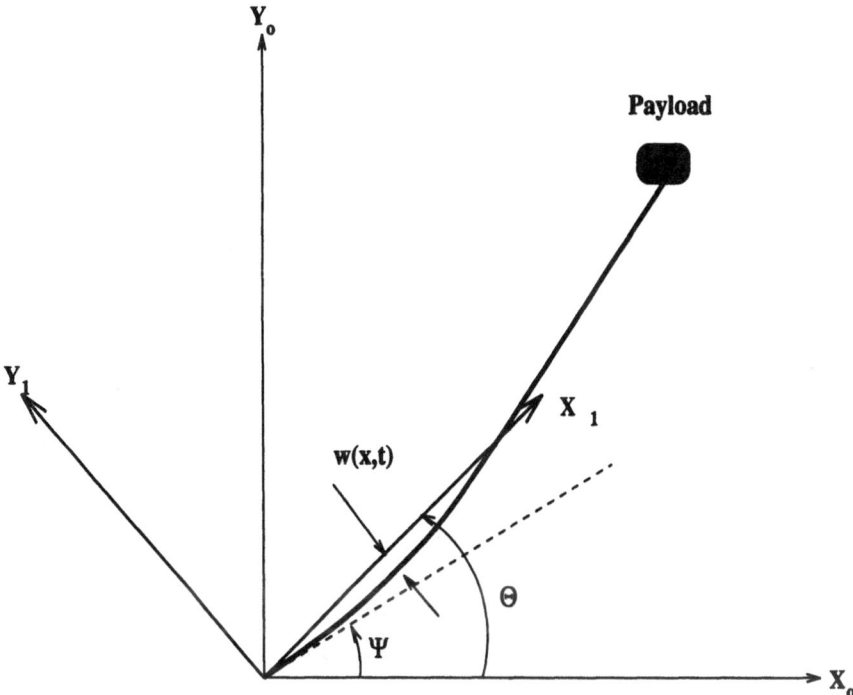

Figure 1.1. A one-link flexible arm.

payload and pinned–free eigenfunctions for mode shapes. With this approximation, the mode shapes form a complete set of orthogonal functions with the rigid body mode shape given by $\phi_0(x) = x$. Using the Lagrangian formulation and taking into account the structural damping of the beam, a set of decoupled differential equations are obtained which can be put in the form of a linear state–space model. Hastings and Book [1] followed a similar approach to modeling. They have experimentally verified that clamped–mass admissible functions (mode shapes) yield better results than other mode shapes such as pinned–free used by Cannon and Schmitz. Two mode shapes have been used for modeling and acceptable agreement with theory has been reported. Wang and Vidyasagar [36] have used clamped–free mode shapes and have theoretically shown that if the number of modes is increased to obtain a more accurate model of the flexible link, the transfer function from torque input to tip position output does not have a well defined relative degree. To alleviate this problem they propose a re–definition of the output as the *reflected tip position* that can be easily measured. Pota and Vidyasagar [11],

and Wang and Vidyasagar [12] further discuss the passivity of the
resulting transfer function and conclude that any strictly passive cas-
cade controller will stabilize the system, and the controllers become
simpler. Bellezza *et al.* [4] have shown that the open loop models
obtained by using clamped–free and pinned–free eigenfunctions are
identical and only differ in the reference frame in which the elastic deflections are measured. It should be noted that the actual boundary
conditions given by (1.5)–(1.8) are neither pinned–free nor clamped–
free. In fact, the boundary conditions are time varying and depend
on the input torque profile. Cetinkunt and Yu [5] have compared
the first three modes of the closed–loop system for pinned-free and
clamped–free mode shapes with the modes obtained from the exact
solution of the Bernoulli–Euler beam equation. For both tip position
proportional–derivative (PD) controller and hub angle PD controller,
they have shown that the predictions of clamped–free mode shapes
are much more accurate than the predictions of pinned–free mode
shape models. Bayo [6] uses Hamilton's principle and a finite element approach to model a single flexible–link arm. One advantage
of this method is that different material properties and boundary
conditions like hubs, tip loads and changes in cross section can be
handled in a simple manner. Usoro *et al.* [7] use a finite-element
Lagrange method to model multi–link flexible robots, but no comparison is made with other schemes and the model is not validated
with experimental evidence. There have been a few other approaches
to modeling multi–link flexible robots. Book [29] introduced a recursive Lagrangian assumed modes method to model a flexible–link
manipulator in three dimensional space. The method is applicable
to revolute–joint robots and gives rise to dynamic equations of the
form given by (1.3). A complete model for a two-link manipulator is
given in [9]. Benati and Morro [10] consider the modeling of a chain
of flexible links by describing the deflection $w(x_i, t)$ of each point on
link i as

$$w(x_i, t) = a_1(t)\phi_1(x_i) + a_2(t)\phi_2(x_i) \qquad (1.9)$$

where ϕ_1, ϕ_2 are the first two eigenfunctions of the clamped–mass
beam with a mass equal to the total mass of the links and payload
after link i. The deflection equation (1.9) can then be related to ψ_i
and δ_i quantities by (see Figure 1.2)

$$w(x_i, \psi_i, \delta_i) = \alpha(x_i, l) + \beta(x_i, l)\delta_i \qquad (1.10)$$

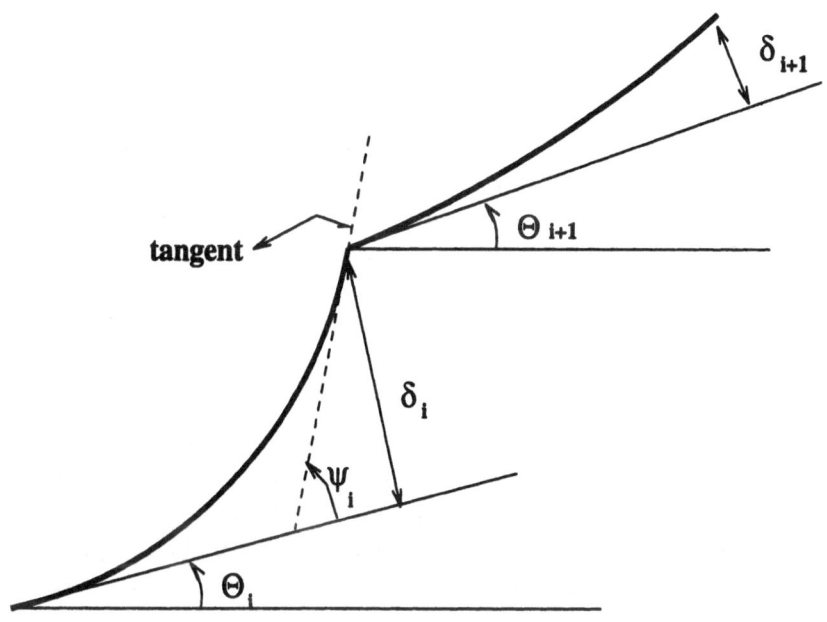

Figure 1.2. Chain of flexible links.

where l is the link length and α, β are known functions. The kinetic and potential energies are then found based on this kinematic description and the Lagrangian method is used to derive the dynamic equations. Yoshikawa *et al.* [13] have modeled each link of a three degree-of-freedom flexible robot with two flexible links using a lumped parameter approximation. The mass of each link is assumed to be negligible and a lumped mass is considered at the tip of each link. The flexibility effects are approximated by lumped torsional and longitudinal springs attached at the end point of each link. These quantities are in terms of the geometry of the links and material properties such as mass density and modulus of elasticity. The modeling is again carried out using the Lagrangian formulation. De Luca *et al.* [14] have conducted experiments on a two–link robot with a flexible forearm. The modeling is similar to that in [9]. They have reported the accuracy of the flexible part of their model based on experimental observations. For a rather severe trajectory, it is concluded that the first vibrational mode is significantly excited and the model with two vibration modes has been reported to yield satisfactory results.

1.3. Control Strategies

A flexible-link manipulator is inherently an underactuated system in the sense that there are fewer control inputs than degrees of freedom. As pointed out by Spong [107], an interesting property that holds for the entire class of underactuated systems is the so-called collocated partial feedback linearization property, which is a consequence of positive definiteness of the inertia matrix. However, collocated control may result in loss of bandwidth and poor behavior at outputs that are of particular interest. In the case of flexible-link manipulators, a collocated control strategy that is based on measuring joint positions and velocities and actuating the joints may result in poor performance due to the oscillatory internal dynamics. Enhancements in bandwidth can be achieved by non-collocated control as described in [2] for a single flexible arm. Thus, a major concern in the design of control schemes for flexible-link manipulators is the robustness of the closed-loop system. From the stand point of system stability, if the system is rigid, the locations of control sensors and actuators would be unimportant. However, when flexure is considered, these locations are of utmost importance because they determine the coupling of the flexural motions to the rigid-body motion to be controlled. It has been known for some time that this coupling can often lead to instability [95]. One way to obtain stable operation is to co-locate sensors and actuators [95]. Although this method may not yield the best performance in terms of the achievable bandwidth, it may result in stable operation. However, co-location may result in undesirable tracking performance if the end-effector of the flexible-link manipulator has to track a desired motion trajectory. To illustrate the problem consider the dynamics of a flexible-link manipulator given by (1.3). Also, let us write the mass matrix in the form

$$M(q,\delta) = \begin{bmatrix} M_{11}(q,\delta) & M_{12}(q,\delta) \\ M_{21}(q,\delta) & M_{22}(q,\delta) \end{bmatrix}.$$

If no flexure is present, the dynamic equations represented by (1.3) reduce to

$$M_{11}(q,0)\ddot{q} + f_1(q,\dot{q}) + h_1(q,0) = u. \tag{1.11}$$

The above dynamics represent the common known rigid dynamics that can be obtained from well known Newton-Euler or Lagrangian methods. A conventional control approach for the above dynamics is the computed-torque method where the control law is given by

$$u = M_{11}(q,0)(\ddot{q}_r + K_d(\dot{q}_r - \dot{q}) + K_p(q_r - q)) + f_1(q,\dot{q}) + h_1(q,0)$$

$$(1.12)$$

where q_r, $\dot{q}_r, \ddot{q}_r$ represents the desired (reference) joint trajectory. Application of this control scheme to (1.12) results in a fully decoupled system with closed-loop behavior given by

$$\ddot{e} + K_d\dot{e} + K_pe = 0 \qquad (1.13)$$

where $e = q_r - q$.

The above control scheme works well for *rigid* manipulators and provides good robustness properties even when uncertainties in the mass matrix and the Coriolis and centrifugal terms are present. However, if flexural effects are significant, a few problems may arise as explained next. First, if the same control scheme is used, the sensor readings at the joint levels also contain components due to structural flexibility. Observation of these effects at the sensor level is known as *observation spillover* [98] that can easily lead to instability due to coupling effects. Similarly, flexural modes can be excited by the control scheme which is only based on rigid-body dynamics. This effect is known as *control spillover*. If the system is not destabilized by this effect, its response may still be unacceptable due to excitation of flexural modes. It should be noted from the second row of the dynamic equations in (1.3) that $\ddot{\delta}$ and $K\delta$ correspond to a marginally stable system with modes on the $j\omega$-axis. Although structural damping is always present in a practical situation, it is usually not enough to damp out vibrational effects in more demanding motion profiles. In fact, the marginally stable poles can easily be driven to the right-half of the s-plane if improper feedback is used. Another major issue in controlling this system is concerned with the ultimate objective of using a robot system. In many applications, we are interested in moving the end-effector payload along a desired trajectory. As we will see later, this is not an easy task due to the non-minimum phase nature of the system. It is clear from the above discussion that in order to obtain robust and acceptable behavior, the flexible modes have to be included both in sensing and actuation. The control scheme should therefore take into account flexural effects which means that the dynamic equations utilized in the control law should embody flexible motions as well. We will study some approaches to the control scheme in the rest of this section.

Most of the research in the area of flexible–link manipulator control has been applied to the case of a single flexible arm. However there is one major critique in this regard. In the multi–link case, the mass matrix contains joint position variables, which introduce considerable nonlinearities in the dynamic equations. For a single flexible link, the mass matrix is only a function of deflection variables which are quadratic type nonlinearities. Thus a single–link may well approximate a linear system, while this is not true in the multi–link case. In spite of this fact, the experiments conducted with a single flexible link provide a basis for multi–link investigations, since both cases suffer from the undesirable non–minimum phase property [66]. This property shows up when the controlled output is the end–effector position. In such a case, a critical situation is encountered when one tries to apply standard inversion techniques for exact trajectory tracking. Any attempt to achieve exact tracking via inversion results in unbounded state trajectories and an unstable closed–loop system. The less difficult problem of end–point stabilization may also become troublesome, although not impossible, because of the non–minimum phase nature. On the other hand, the tracking of joint trajectories can always be obtained in a stable fashion in the presence of link flexibilities. This may of course yield unacceptable tip position errors. We will discuss the problem in more detail in Section 1.5.

Of the early experimental work in this area, the work of Cannon and Schmitz [2], Hastings and Book [1], among others, was aimed at the end–point regulation problem. Output re–definition may be a key to achieving smaller tracking errors. To this end, Wang and Vidyasagar [36] introduced the reflected–tip position as a new output for a single flexible link. De Luca and Lanari [37] studied the regions of sensor and actuator locations for achieving minimum phase property for a single flexible link. Wang and Vidyasagar [56] showed that the nonlinear flexible–link system is not in general input–state feedback linearizable; however the system is locally input–output linearizable. Nemir et al. [64] introduced the pseudo–link concept but have not addressed the non–minimum phase issue associated with the tip output. Other approaches have been proposed to deal with the exact tracking problem. Bayo [6], and Kwon and Book [24] introduced noncausal controllers for the purpose of exact trajectory tracking. However their approaches typically require heavy computation and have been restricted to the linearized single–link case. Based on the concept of pole assignment in linear systems, transmission zero assignment was introduced by Patel and Misra [41] and applied by Geniele *et al.* [54] to a single–link flexible manipulator. Here the basic idea is to add a feedthrough block to the plant so that the zeros of the new system are at prescribed locations in the left half–plane. It is then possible to use output feedback and ensure that the closed–loop system poles are at good locations in the left half–plane. The work of Hashtrudi–Zaad and Khorasani [43] which is based on an integral manifold approach may also be interpreted as a form of output re–definition. In this work, new *fast* and *slow* outputs are defined and the original tracking problem is reduced to tracking the *slow* output and stabilizing the *fast* dynamics. Input–shaping control was implemented, among others, by Hillsley and Yurkovich [25], Tzes and Yurkovich [26], and Khorrami [27]. This approach essentially involves the convolution of a sequence of impulses with the reference inputs to suppress the tip–position vibrations. This can be accomplished by coloring the input such that no energy is injected around the flexible modes, or by filtering out the frequencies around the flexible modes using notch filters. The validity of such methods depends on the exact knowledge of the flexible structure dynamics. Also such methods are open loop in essence.

The nonlinear approach to the design of controllers has also been addressed, although to a lesser extent by some authors, e.g., [8], [16], [48], [49],[56]. In this regard, the approach based on singular perturbation theory [81], [44] has been attractive due to the two–time–scale nature of the system dynamics. To this end, Siciliano and Book [8] furnished a singular perturbation model for the case of multi–link manipulators which follows a similar approach in terms of modeling to that introduced by Khorasani and Spong [46] for the case of flexible–joint manipulators. Their control strategy is then based on stabilizing the fast dynamics and tracking the joint trajectories. The same strategy is also addressed in [16], [48], [49]. Mostly these researchers have taken joint positions as outputs to avoid the problems due to the non–minimum phase nature of the plant. A comparison is made experimentally between some of these methods by Aoustin *et al.* [50]. However, taking joint positions as the outputs has the drawback of large tip position errors, especially when the singular perturbation parameter is not small enough.

Wang and Vidyasagar [56] studied the input–state feedback linearization problem and showed that the system is not in general linearizable, however it is input–output linearizable. The concept of input–output linearization in nonlinear systems theory is essentially based on the works of Hirschorn [18] and Byrnes and Isidori [19]. To this end, the tip positions cannot be selected as the outputs due to the instability of the unobservable dynamics associated with such choice of outputs. Thus the definition of another output may seem unavoidable. Such outputs should naturally be selected such that small enough tracking errors are achieved [37], [38].

The application of *smart* materials, such as piezoelectric ceramics, in vibration control has recently become more attractive, as they can provide convenient actuation, are inexpensive, and can also be used as sensors. Most of the current research has, however, been conducted in the area of flexible structures having linear dynamic behavior [108], [109], [110]. Although these studies have provided valuable guidelines, the control problems due to unstable internal dynamics, nonlinear dynamics, and high order dynamics, are still open with regard to usage of these actuating elements, for example, in the nonlinear multi-link flexible robots.

Recently, intelligent control methods have been applied to flexible–link manipulators. Intelligent control is the discipline in which control algorithms are developed by emulating certain characteristics of biological systems. In this regard, Moudgal *et al.* [39] have implemented a *fuzzy* supervisory control law for vibration damping of a flexible two–link manipulator. In [93], neural network-based controllers are tested, by simulation and experiment, on a single-link flexible arm. However, an unbiased and accurate comparison of intelligent and classical control strategies is not yet available in the literature. Therefore more benchmark comparisons are needed before one can come to any conclusion regarding the superiority of intelligent, classical, or a combination of both methods.

1.4. Mechanical Design

Improving the plant characteristics for achieving a well–behaved system for the purpose of control design has been pursued in systems and control engineering for many years now. As an example, for an aircraft to be open–loop stable, the center of mass has to be ahead of the center of pressure. Thus, one desirable aspect of aircraft design is to guarantee that this condition is satisfied. The same philosophy can also be applied to flexible structure robots. Earlier, we mentioned some of the problems in connection with controlling a flexible structure robot. A natural question that may arise is: Can we design the robot such that these problems can be resolved or at least minimized so that we can use simpler control schemes? Obviously, such designs should not compromise other requirements for which the robot has become flexible (i.e., total inertia of the robot). While some characteristics that are inherent to the flexible-link system cannot be diminished, it would be interesting to know if they can be improved. Intuitively, the source of problems related to control is the flexibility itself. Reducing flexibility is a qualitative measure that can be translated into quantitative measures for the purpose of mechanical design. For example, if the first structural natural frequency of the manipulator system is increased, while the inertia is kept constant, it is expected that the system behavior will be improved. A qualitative justification for this is that the first natural frequency for a *rigid* system is very large. As was mentioned previously, by mechanical design we mean constructing the optimum shape of the links of a manipulator such that some desirable features are achieved. Parallel to this line of research, the use of advanced materials in arm construction should not be overlooked. One example is using distributed actuators and sensors as described in [28]. As pointed out by Asada, *et al.* [21], the majority of flexible manipulators that have been addressed in the literature have a simple structure consisting of beams with uniform mass and stiffness distribution. While the simplified beams allow for analytic modeling and theoretical treatment, the arm construction is unrealistically primitive and its dynamic performance is severely limited. Compensating for the poor dynamics merely by control may require a lot of control energy and heavy computation. Therefore, alteration of plant dynamics to achieve less stringent control strategies can be pursued in this regard.

The design process regarding arm shape design to achieve properties such as low mass and moments of inertia and high natural frequencies generally boils down to the solution of an optimization problem. Asada *et al.* [21] have obtained the optimum torque application point and structural shape for a single flexible arm. The torque application point affects system zeros and the structural shape mostly affects the modal frequencies. The experimental results obtained show a twenty five percent increase in the lowest natural frequency and the plant requires less stringent control strategy due to a robust allocation of the torque transmission mechanism. A few other papers have recently appeared in this area (e.g. [22], [82]).

1.5. Some Related Topics

In this section, we review some basic concepts and definitions that are related to the control of flexible–link manipulators. The challenge of control in these manipulators stems from several sources. First, the dynamic equations are highly coupled and nonlinear. Second, these equations are *stiff* due to the time–scale separation of the *slow* rigid modes and the *fast* flexible modes. Third, the presence of right half–plane transmission zeros in a non–colocated sensor-actuator configuration imposes limitations on the control problem, both in trajectory tracking and set–point regulation. These factors will create problems in control design, analysis and simulation as we will briefly discuss in this chapter.

The study of control for these manipulators is also stimulating if one considers similarities with other control problems. We will give a few examples from other applications where the controller has to cope with plant nonlinearities and unstable zero–dynamics with essentially similar characteristics.

1.5.1. Regulation and Tracking

The task of every control problem can generally be divided into two categories: Regulation (or stabilization) and tracking (or servoing). In the regulation problem, one is concerned with devising the control law such that the system states are driven to a desired final equilibrium point and stabilized around that point. In the tracking problem, one is faced with devising a controller (tracker) such that the system output tracks a given time–varying trajectory. Some examples of regulation problems are: Temperature control of refrigerators, AC and DC voltage regulators, and robot joint position control. Examples of tracking problems can be found in tracking antennas, trajectory control of robots for performing specific tasks, and control of mobile robots. The formal definitions of the above control problems can be stated as follows [66]

Regulation Problem: *Given a nonlinear dynamic system described by*

$$\dot{x} = f(x, u, t) \tag{1.14}$$

where x is the $n \times 1$ state vector, u is the $m \times 1$ input vector, and t is the time variable, find a control law u such that, starting from anywhere in a region in $\Omega \subset R^n$, the state tends to zero as $t \to \infty$.

Similarly,

Tracking Problem: *Given a nonlinear dynamic system and its output vector described by*

$$\begin{aligned} \dot{x} &= f(x, u, t) \\ y &= h(x) \end{aligned} \tag{1.15}$$

and a desired output trajectory $y_d(t)$, find a control law for the input u such that, starting from an initial state in a region $\Omega \subset R^n$, the tracking error $y(t) - y_d(t)$ approaches zero while the whole state x remains bounded.

The control input u in the above definitions may be either called *static* if it depends on the measurements of the signals directly, or *dynamic* if it depends on the measurements through a set of differential equations. Tracking problems are generally more difficult to solve than regulation problems. One reason is that, in the tracking problem, the controller has to drive the outputs close to the desired trajectories while maintaining stability of the whole state of the system. On the other hand, regulation problems can be regarded as special cases of tracking problems when the desired trajectory is constant with time.

In this monograph we are concerned with the tip–position tracking problem of flexible–link manipulators. To get more insight into the nature of this problem we will first review some concepts such as internal dynamics, zero–dynamics, and non–minimum phase characteristic in the nonlinear framework. Towards this end let us consider a class of square nonlinear systems given by (1.15) that are linear in the input u (affine systems), i.e.,

$$\begin{aligned} \dot{x} &= f(x) + g(x)u \\ y &= h(x). \end{aligned} \tag{1.16}$$

Let $x \in R^n$, $u \in R^p$, $y \in R^p$. Further, assume that the column vector fields $f(x)$, $g(x)$, and the function $h(x)$ are smooth on an open set of R^n. We are interested in tracking the output vector y while keeping all other states reasonably bounded. In this regard, a representation of (1.16) in terms of the output vector and its time derivatives would prove helpful. Towards this end, assume that the system (1.16) has well defined vector relative degree ([31])

$$r = [r_1,\ r_2,\ ...,\ r_p] \tag{1.17}$$

in a neighborhood of the origin $x = 0$. Also assume that $f(0) = 0$ and $h(0) = 0$. The assumption of a well defined relative degree implies that if we successively differentiate $y_i(t)$ with respect to t, then some component $u_j(t)$ of the input $u(t)$ appears for the first time at the r_ith derivative of y_i.

Through state–dependent coordinate and input transformations, ([31]), we may input–output linearize the plant (1.16) so that it takes the equivalent representation

$$
\begin{aligned}
\dot{\xi}_i^j &= \xi_{i+1}^j, \quad i \in \{1,\ ...,\ p\},\ j \in \{1,\ ...,\ r_i - 1\} \\
\dot{\xi} &= u_i, \quad i \in \{1,\ ...,\ p\} \\
\dot{\eta} &= \alpha(\xi, \eta) + \beta(\xi, \eta)u \\
y_i &= \xi_1^i, \quad i \in \{1,\ ...,\ p\}
\end{aligned}
\tag{1.18}
$$
$$\tag{1.19}$$

where

$$
\xi := \Psi(x) =
\begin{bmatrix}
\xi_1^1 \\
\xi_2^1 \\
\vdots \\
\xi_{r_1}^1 \\
\xi_1^2 \\
\vdots \\
\xi_{r_2}^2 \\
\xi_1^p \\
\vdots \\
\xi_{r_p}^p
\end{bmatrix}
=
\begin{bmatrix}
y_1 \\
\dot{y}_1 \\
\vdots \\
y_1^{(r_1-1)} \\
y_2 \\
\vdots \\
y_2^{(r_2-1)} \\
y_p \\
\vdots \\
y_p^{(r_p-1)}
\end{bmatrix}.
\tag{1.20}
$$

The ξ vector represents part of the dynamics in the form of p integrator chains, each of length r_i, $i \in \{1,\ ...,\ p\}$, and $\eta^T = [\eta_1, \cdots, \eta_{n-p}]$

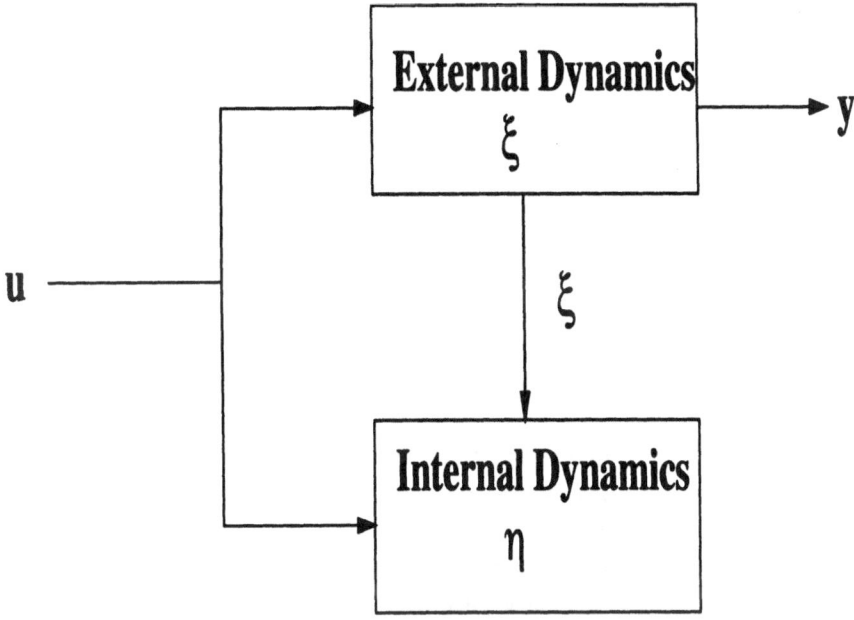

Figure 1.3. Representation of the system in internal and external dynamics.

$(\eta_i = \phi_i(x),\; i = 1,\; \cdots,\; n - p)$, are smooth functions of x such that the state transformation $(\xi, \eta) = \Psi(x) := (\psi(x), \phi(x))$ is a (local) diffeomorphism at the origin.

The structure of (1.18)–(1.19) is illustrated in Figure 1.3. The dynamics of η are referred to as *internal* dynamics. These dynamics are obtained from (1.19) with ξ regarded as an exogenous time-dependent function. Now the *zero–dynamics* of the system in Figure 1.3 are defined as the internal dynamics of the system when the inputs act such that the output is identically zero ($u \equiv 0,\; \xi \equiv 0$). In the representation thus obtained, the zero–dynamics are given by

$$\dot{\eta} = \alpha(0, \eta). \tag{1.21}$$

If the zero–dynamics of (1.21) are asymptotically stable then the original system is a minimum–phase system. Otherwise the system is nonmimimum–phase. These terms are adapted from linear system theory. If a transfer function $H(s)$ of a linear system has a zero in the right half of the complex plane, the transfer function, when evaluated for s going along the imaginary axis from $-j\infty$ to $+j\infty$, undergoes a change in phase which is greater, for the same magnitude, than if that zero were replaced by its left–half plane mirror image; hence the name non–minimum phase. In linear systems the minimum–phase characteristic can be checked in different ways. To this end the transmission zeros of a controllable and observable nth–order m input, m output linear system of the form

$$
\begin{aligned}
\dot{x} &= Ax + Bu \\
y &= Cx
\end{aligned}
$$
(1.22)

are defined as those values of λ for which

$$
rank \begin{bmatrix} A - \lambda I_n & B \\ C & 0 \end{bmatrix} < n + m.
$$
(1.23)

Similarly, if CB is invertible the transmission zeros are the finite eigenvalues of $A + gBC$ as $g \to \infty$. The latter definition has an interesting interpretation: When static output feedback $(gI_{m \times m})$ is used the poles of the closed–loop system are attracted towards the transmission zeros as the feedback gain g is increased. Thus if the system has right–half plane transmission zeros (non–minimum phase) the closed–loop system can become unstable under static output feedback.

1.5.2. Some Examples of Non–minimum Phase Systems

In the rest of this section, we will consider some examples in other areas with non–minimum phase behavior. Most of these examples are from underactuated mechanical systems which have some similarities to the underactuated flexible–link system.

The Acrobot

The acrobatic–robot [87] or *acrobot* (see Figure 1.4) is a highly simplified model of a human gymnast performing on a single parallel bar. By swinging his/her legs (a rotation at the hip) the gymnast is able to bring himself/herself into a completely inverted position with the feet pointing upwards and the center of mass above the bar. The acrobot consists of a simple two–link manipulator operating in a vertical plane. The first joint (corresponding to the gymnast's hand sliding freely on the bar) is free to rotate. A motor is mounted at the second joint, between the links, to provide torque input to the system. This corresponds to the gymnast's ability to generate torques at the hip. The acrobot is also a good model for under–actuated

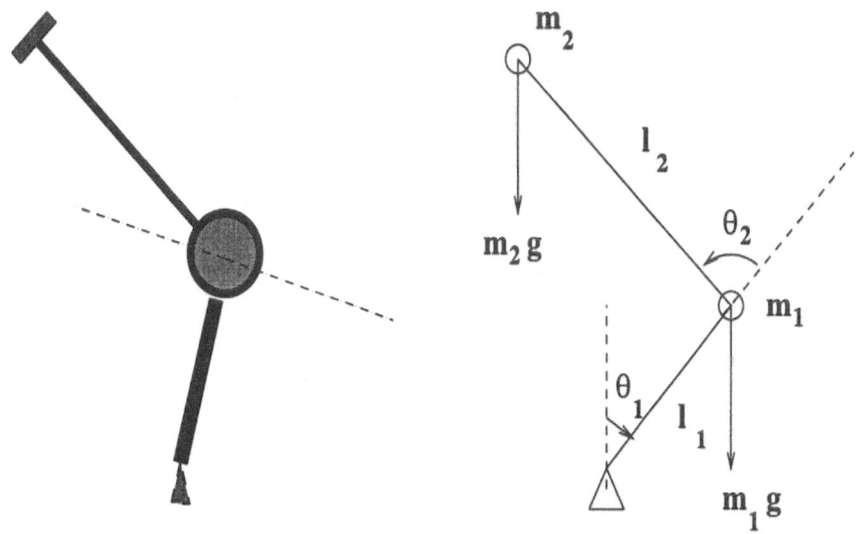

Figure 1.4. Acrobot: An acrobatic robot.

mechanical systems such as unicycles and walking machines, where *balance* must be maintained while trying to accomplish the assigned task. The dynamics of the acrobot can be obtained using the Lagrangian formulation, and take the general form

$$M(\theta)\ddot{\theta} + C(\theta,\dot{\theta}) + G(\theta) = \begin{pmatrix} 0 \\ \tau \end{pmatrix} \qquad (1.24)$$

where $\theta = (\theta_1, \theta_2)$ are the joint angles, M is the positive–definite inertia matrix, C contains the Coriolis and centrifugal forces, G contains the effects of gravity, and τ is the torque applied between the first and second links.

Now modeling the acrobot shown in Figure 1.4 with $m_1 = m_2 = 8kg$, $l_1 = 0.5m$, $l_2 = 1m$, and $g = 10m/s^2$ results in

$$M(\theta) = \begin{bmatrix} 12 + 8cos\theta_2 & 8 + 4cos\theta_2 \\ 8 + 4cos\theta_2 & 8 \end{bmatrix}$$

$$C(\theta,\dot{\theta}) = \begin{bmatrix} -4\dot{\theta}_2(2\dot{\theta}_1 + \dot{\theta}_2)sin\theta_2 \\ 4\dot{\theta}_1^2 sin\theta_2 \end{bmatrix}$$

$$G(\theta) = \begin{bmatrix} -80(sin\theta_1 + sin(\theta_1 + \theta_2)) \\ -80sin(\theta_1 + \theta_2) \end{bmatrix}. \qquad (1.25)$$

Consider the equilibrium point corresponding to $\theta_1 = \theta_2 = 0$, and choose the desired output to correspond to θ_1, i.e., $y = \theta_1$. Then the zero–dynamics of this system can be obtained by finding an input that restricts the system output identically to zero and by considering the resulting internal dynamics. To this end setting y, $\dot{y}$, $\ddot{y}$ to zero yields from the first equation in (1.24) the zero–dynamics

$$(8 + 4cos\theta_2)\ddot{\theta}_2 - 4\dot{\theta}_2^2 sin\theta_2 - 80sin\theta_2 = 0 \qquad (1.26)$$

and the corresponding input is obtained from the second equation in (1.24), i.e.,

$$8\ddot{\theta}_2 - 80sin\theta_2 = u. \qquad (1.27)$$

Around the equilibrium point $\theta_2 = 0$, $\dot{\theta}_2 = 0$, the dynamics given by (1.26) can be linearized to yield

$$12\ddot{\theta}_2 - 80\theta_2 = 0. \qquad (1.28)$$

Thus the zero–dynamics are *unstable* at this equilibrium point. Physically these dynamics correspond to the dynamics of the second link (an inverted pendulum) when θ_1 (output) is restricted to zero. Moreover, equation (1.27) illustrates how the control input is affected by the *unstable* zero–dynamics in order to maintain balance at $\theta_1 = \theta_2 = 0$. It can be shown that taking θ_2 as the output will also result in unstable zero–dynamics.

PVTOL Aircraft

We consider the planar vertical takeoff and landing (PVTOL) aircraft given in [88]. Referring to Figure 1.5 the aim is to control the position of the aircraft. Thus the outputs to be controlled are x and y. Simplified equations of motion for this system are

$$
\begin{aligned}
\ddot{x} &= -u_1 sin\theta + \epsilon u_2 cos\theta \\
\ddot{y} &= u_1 cos\theta + \epsilon u_2 sin\theta - 1 \\
\ddot{\theta} &= u_2
\end{aligned}
\qquad (1.29)
$$

where "1" is the gravitational acceleration and ϵ is the (small) coefficient giving the coupling between the rolling moment and the lateral acceleration of the aircraft. The control inputs u_1, u_2, are the thrust (directed out from the bottom of the aircraft) and the rolling moment. The non–minimum phase characteristic of the aircraft is a

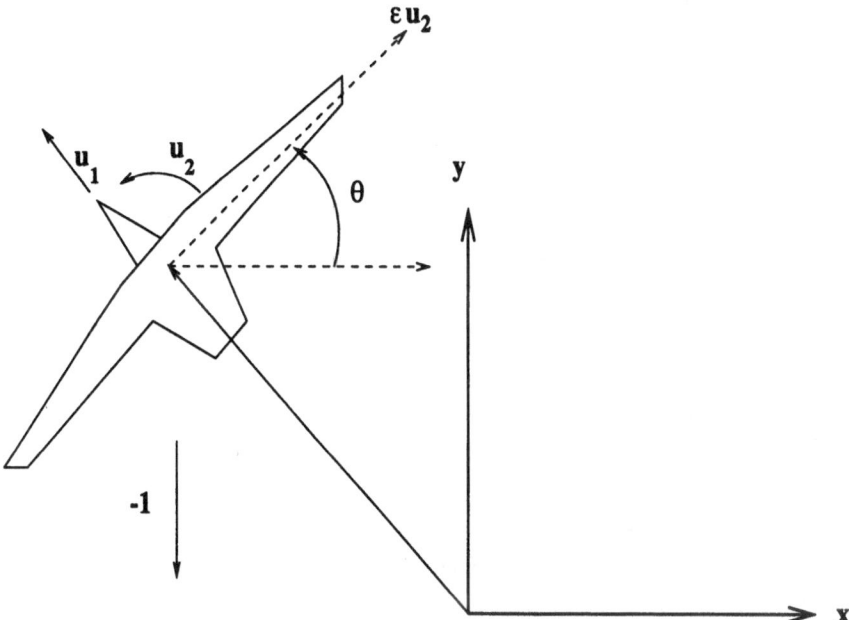

Figure 1.5. A planar vertical take-off and landing (PVTOL) aircraft.

result of small body forces that are produced in the process of generating body moments. Here since the roll moment reaction jets create a force that is not perpendicular to the lateral axis of the aircraft, the production of a positive rolling moment (to the pilot's right) will also produce a slight acceleration of the aircraft to the left. This phenomenon makes the aircraft non–minimum phase. Thus taking x and y as the outputs, the zero dynamics are obtained by first differentiating each output until at least one input appears and then setting each output identically to zero. This will lead to the following equivalent representation

$$
\begin{aligned}
\ddot{x} &= v_1 \\
\ddot{y} &= v_2 \\
\ddot{\theta} &= \frac{1}{\epsilon}(\sin\theta + v_1\cos\theta + v_2\sin\theta).
\end{aligned}
\tag{1.30}
$$

Thus the zero–dynamics are given by

$$
\ddot{\theta} = \frac{1}{\epsilon}\sin\theta.
\tag{1.31}
$$

The above equation is reminiscent of the dynamics of an undamped pendulum which is not asymptotically stable. This undamped behavior may produce undesirable aircraft response. For example if y is to be kept at zero by v_2 and x is forced to track a smooth trajectory by v_1, the aircraft will acquire a pendulum–like motion which may not be desirable.

Another non–minimum phase characteristic in aircraft dynamics is the response of aircraft altitude to the deflection of the elevator angle [66]. This will show itself in an initial downward motion when an attempt is made to move upward. Of course, such non–minimum phase behavior is important for the pilot to know, especially when flying at low altitudes.

Flexible–link Manipulators

To illustrate the non–minimum phase characteristic in flexible–link systems, we consider a single–link flexible arm as shown in Figure 1.1. A linearized dynamic model of this arm can be found by using the method of assumed modes and the Lagrangian formulation. When a single flexural mode is used the following dynamics result

$$
\begin{aligned}
m_{11}\ddot{\theta} + m_{12}\ddot{\delta} &= u \\
m_{12}\ddot{\theta} + m_{22}\ddot{\delta} + k\delta + d\dot{\delta} &= 0
\end{aligned}
\tag{1.32}
$$

where m_{ij}, $i,\ j \in \{1, 2\}$, are components of the mass matrix given for zero payload by

$$
m_{11} = \rho A l^3 / 3 + J_h, \quad m_{12} = \int_0^l \phi(x)x\rho A dx, \quad m_{22} = \rho A \tag{1.33}
$$

where A is the link's cross sectional-area, ρ is its mass density, $\phi(x)$ is the modal shape function, and k is the stiffness coefficient of the beam given by

$$
k = EI \int_0^l (\frac{d^2\phi}{dx^2})^2 dx. \tag{1.34}
$$

In (1.34), E is the link modulus of elasticity and I is the cross sectional area moment of inertia. Also in (1.32), d is the damping coefficient of the flexural mode, and u is the input torque. Defining the output y as the tip position and $\phi_e = \phi(l)$, we have

$$
y = \theta + \phi_e \delta. \tag{1.35}
$$

As before the zero–dynamics can be found by setting y identically to zero, which yields

$$u = (m_{12} - m_{11}\phi_e)\ddot{\delta} \qquad (1.36)$$

$$0 = (m_{22} - m_{12}\phi_e)\ddot{\delta} + d\dot{\delta} + k\delta. \qquad (1.37)$$

The zero–dynamics represented by (1.37) are generally unstable since $m_{22} - m_{12}\phi_e$ can be negative if the length of the arm is large enough. Note that by using (1.33) the latter term is given by $\rho A(1 - \phi_e \int_0^l x\phi(x)dx)$. Assuming a clamped–free mode shape for $\phi(x)$, and noting that $\phi(x)$ is in the first quadrant for $0 < x < l$, it can be concluded that the term $x\phi(x)dx$ is always positive. Therefore, if l is large enough the first coefficient in (1.37) is negative, i.e., unstable zero–dynamics. The non-minimum phase condition in this case is a result of the non–colocated sensor and actuator positions. The system is under–actuated as in the previous examples and the input torque affects the tip position through the flexural variable δ and rigid body mode θ. From the control point of view, the zero–dynamics address an important question: Is there any control input that can identically regulate y to zero? The unstable zero–dynamics given by (1.37) imply that the internal states δ, $\dot{\delta}$ will be unbounded if the initial states are different from zero. This will require an unbounded input from (1.36) which is not desirable. However, relaxing the control goal of identically zeroing the output, it is possible to have sufficiently small bounded output while the internal states are bounded, even when the system is non-minimum phase.

Macro-micro Manipulator Systems

Initially, macro-micro (abbreviated as M-m) manipulators were proposed for long reach tasks requiring speed and precision. Rigid M-m manipulators were studied by Khatib [96]. However, for long-reach tasks the flexibility of the M-part becomes significant. Chiang *et al.* [97] have studied the control problem of a system composed of a very flexible forearm carrying a rigid wrist that is actuated by a dc motor. The fundamental result of this work is that it illustrates how the geometric characteristics of the m-manipulator can affect the system zeros. The dynamics of the macro-micro manipulator system described here are essentially similar to those of a flexible-link manipulator system. In fact, these manipulators belong to a special class of flexible-link manipulators where the last links corresponding to the micro manipulator are rigid. In many applications, the macro manipulator joints are locked, and the system can be considered as a non-actuated flexible link (base) with a rigid manipulator mounted on its end. For example, in power-line maintenance, a large manipulator carries a small robot near the power line. The large manipulator joints are then locked and the smaller robot can do more exact manipulation such as handling a heavy transformer. In nuclear waste disposal, where safety is a critical issue, the large robot is locked and fine manipulation is performed by the smaller robot. In such cases, the system can be considered as a rigid manipulator operating on a flexible platform. A simplified model of such a system is represented in Figure 1.6. This system is also non-minimum phase if the outputs are taken as the end-effector position and orientation. There are some other interesting applications in space that resemble the above system. Consider a space robot mounted on a spacecraft. The motion of the robot can cause reaction forces that tend to move the spacecraft as well. One way to nullify these effects is to use reaction jets on the spacecraft to produce counterbalancing forces and moments. This requires energy consumption due to reaction jets for robot manipulation. Another approach is to let the robot itself compensate for the effects of its motion and the resulting base motion, i.e., a *free-flying* robot. In such a case the robot should move such that it induces a minimum amount of motion in the base.

1.6.   Outline of the Monograph

This monograph focuses on three main steps in engineering practice: Theoretical development, software simulation, and practical implementation. Here we are specifically dealing with the trajectory tracking control of flexible multi-link manipulators and are interested in achieving sufficiently small tip-position tracking errors while maintaining closed-loop system stability.

The control problem is considered to be a difficult one for several reasons: The non-minimum phase characteristic of the plant, the nonlinear dynamics of the manipulator, and the ill-conditioned dynamics that result from the time-scale separation of rigid and flexible modes. It is now more than a decade since several researchers first worked on different aspects of the control and design of such ma-

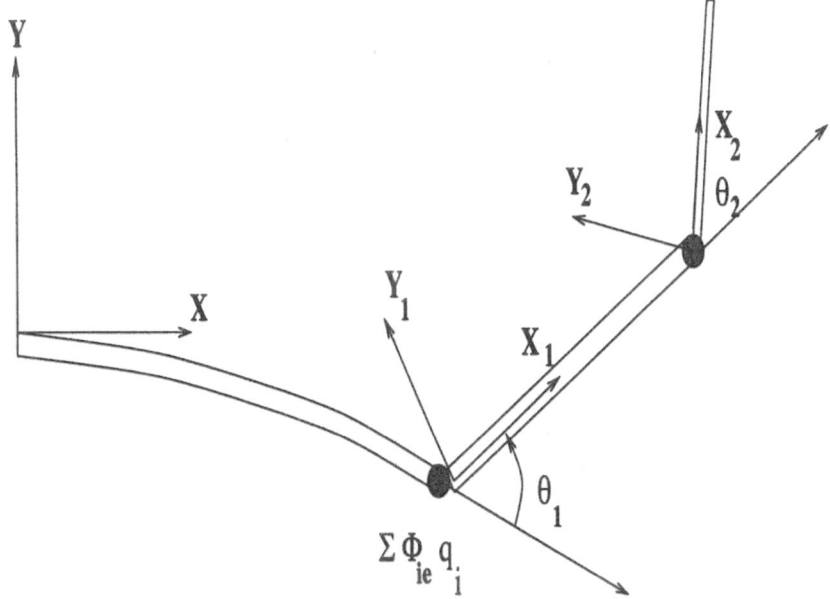

Figure 1.6. A boom-mounted two-link manipulator.

nipulators. In this respect, the research described in this monograph can be summarized as follows:

1.6.1. Tracking Control Using Integral Manifolds

Development of a nonlinear control strategy is considered for approximate tip–position tracking of a class of flexible multi–link manipulators based on the concept of integral manifolds and singular perturbation theory. The development is along the lines stated in [43] which is applicable to the linear dynamics of a single–link flexible arm. Our development is based on the more appropriate *nonlinear* framework, and is applicable to a class of multi–link flexible manipulators. The results are stated in Theorem 2.1 which furnishes the conditions under which small tracking errors and closed–loop system stability are guaranteed. From a practical point of view, a major advantage of the proposed strategy is that the only measurements required are the tip positions, joint positions, and joint velocities. This topic is addressed in Chapter 2.

1.6.2. Tracking Control by Output Redefinition and Input–output Decoupling

Decoupling control is essentially based on the developments described in [18] and [19]. Its application to flexible–link manipulators has appeared in [79] and [38]. To ensure that the internal dynamics remain bounded, we have modified both the controlled outputs and the control inputs such that boundedness of the internal dynamics is guaranteed. The results are summarized in Theorem 3.1 which indicates the conditions for achieving closed–loop system stability with this control strategy. Again, the control strategy is developed in a general context for a class of multi–link flexible manipulators. This topic is addressed in Chapter 3. The strategy is further expanded in Chapter 5, using the concept of sliding surfaces in variable structure control. It is shown that a more robust performance is achieved in the face of considerable parametric uncertainties. The results are summarized in Theorem 5.1.

1.6.3. Observation Strategy for Flexural Rates

Many advanced control strategies require knowledge of flexible modes as well as their time derivatives (flexural rates). The flexible modes can be measured by economic sensors such as strain gauges, but the measurement of flexural rates is rather inconvenient and prone to errors. Therefore, an observer is proposed to estimate these variables. The observation scheme is proposed in a general framework and can be applied to find the flexural rates if joint positions, joint velocities, and flexible modes are available. This topic is addressed in Chapter 4. The observation scheme is also incorporated in the control strategy outlined in item 2 above and the conditions for achieving closed–loop system stability are obtained. The results are summarized in Theorem 4.1.

1.6.4. Structure Design

Improving the plant characteristics to achieve a more well behaved system for the purpose of control has been a usual trend in systems and control engineering practice. As an example, for the aircraft to be open–loop stable, the center of mass has to be ahead of the center of pressure. Thus, one aspect of aircraft design is to achieve such a condition. The same philosophy can be applied to flexible structure robots. The design process then boils down to the solution of an optimization problem to achieve low inertia arms with high structural natural frequencies [21], [22], [82]. Although increasing the structural natural frequencies will help to improve structural properties, it is not necessarily enough to achieve a more robust control. In this regard, an optimization procedure is presented in Chapter 6 that utilizes controllability and observability properties of the flexible-link system in designing the arm.

1.6.5.   Experimental Evaluation

To evaluate the performance of the controllers outlined in parts 1–3 above a two–link flexible manipulator was built with the first link rigid and the second link flexible. This setup has two significant features that highlight two main characteristics of flexible manipulators: Nonlinear dynamics and non–minimum phase behavior. The instrumentation, wiring, layout design, analog signal conditioning, and interfacing with the computational engine were carried out in the first step. In the second step, the control algorithms were coded in *C* and tailored for execution in a real–time environment. More details on the experimental results are given in Chapters 2 and 4.

2.

Tracking Control by Integral Manifolds

In this chapter a nonlinear control strategy for tip position trajectory tracking of a class of structurally flexible multi–link manipulators is developed. Using the concept of integral manifolds and singular perturbation theory, the full–order flexible system is decomposed into *corrected slow* and *fast* subsystems. The tip position vector is similarly partitioned into corrected *slow* and *fast* outputs. To ensure an asymptotic tracking capability, the corrected *slow* subsystem is augmented by a dynamical controller in such a way that the resulting closed–loop zero dynamics are linear and asymptotically stable. The tracking problem is then *re–defined* as tracking the *slow* output and stabilizing the corrected *fast* subsystem by using *dynamic output* feedback. Consequently, it is possible to show that the tip position tracking errors converge to a residual set of $O(\varepsilon^2)$, where ε is the singular perturbation parameter. A major advantage of the proposed strategy is that the only measurements required are the tip positions, joint positions, and joint velocities. Experimental results for a single–link arm are also presented and compared with the case when the slow control is designed based on the rigid–body model of the manipulator.

2.1. Introduction

As it was discussed in Chapter 1 the control of structurally flexible manipulators is hampered by their non–minimum phase characteristic. For a causal controller, this characteristic hinders perfect asymptotic tracking of the desired tip position trajectories with bounded control inputs. In this regard, the approach based on singular perturbation theory [81], [44], is attractive due to the two–time–scale nature of the system dynamics.

In this chapter, we address the problem of tip position tracking of flexible multi–link manipulators with the same design philosophy as in [43],[90] but by taking into account the nonlinear characteristics of the plant. Most of the standard singular perturbation results that are applied to flexible–link manipulators in the literature exclude high performance light–weight manipulators, since a reduced–order *rigid body* equivalence of the flexible manipulator has limited use and application. However, the integral manifold approach in [43] and [45], [46] facilitates the inclusion of the effects of higher frequency flexible modes into the corrected models. The methodology proposed in this chapter is tested by simulations on a two–link flexible manipulator, and experimentally on a single–link flexible arm. The new strategy allows for smaller tip position tracking errors, and its implementation does not require any measurement of rates of change of deflection variables with time, as these variables are not generally conveniently measurable.

The organization of this chapter is as follows. In Section 2.2, the concept of integral manifolds is used to decompose the dynamics of the full–order flexible system into reduced–order corrected *slow* and *fast* subsystems up to $O(\varepsilon^3)$, where ε is the singular perturbation parameter representing the elasticity of the arm. In Section 2.3, the control laws for the reduced order subsystems are designed. The main objective of the control problem is to achieve asymptotic stability of the fast subsystem and guarantee the tracking of the slow subsystem outputs by using only tip positions, joint positions and velocities. Towards this end, *dynamical* output feedback controllers are constructed for the corrected *slow* and *fast* subsystems. It is shown that in the resulting closed–loop system, the tip positions track desired reference trajectories to $O(\varepsilon^2)$. In Section 2.3, simulation results are also presented for a two–link flexible manipulator. Experimental results for a single–link arm are presented in Section

2.4 and compared with the case when the *slow* control is based on the rigid–body model of the manipulator. It is shown that improved tip position tracking performance is achievable by using the proposed scheme. Finally, conclusions are given in Section 2.5.

2.2. Model Reduction Using Integral Manifolds

In this section, an order reduction of the dynamic equations of flexible–link manipulators is given by using the concept of integral manifolds. A composite control strategy [44] is assumed in which the controller is comprised of *slow* and *fast* terms. The integral manifolds method, in the context of composite control, has been applied to flexible–joint manipulators in [45] and [46] and to flexible–link manipulators in [43]. Such control laws are referred to as corrective control laws because the slow control component contains corrective terms added to the term designed for the rigid model. In the following, we briefly establish the singularly perturbed model of the flexible–link manipulator that is considered in the sequel.

Consider the dynamic equations of a flexible–link robotic manipulator given by ([8])

$$M(q,\delta)\begin{bmatrix} \ddot{q} \\ \ddot{\delta} \end{bmatrix} + \begin{bmatrix} f_1(q,\dot{q}) \\ f_2(q,\dot{q}) \end{bmatrix} + \begin{bmatrix} g_1(q,\dot{q},\delta,\dot{\delta}) \\ g_2(q,\dot{q},\delta,\dot{\delta}) \end{bmatrix} + \begin{bmatrix} O & O \\ O & K \end{bmatrix}\begin{bmatrix} q \\ \delta \end{bmatrix}$$

$$= \begin{bmatrix} u \\ 0 \end{bmatrix} \qquad (2.1)$$

or equivalently

$$\begin{aligned}
\ddot{q} &= -H_{11}(q,\delta)(f_1(q,\dot{q}) + g_1(q,\dot{q},\delta,\dot{\delta})) \\
&\quad - H_{12}(q,\delta)(f_2(q,\dot{q}) + g_2(q,\dot{q},\delta,\dot{\delta})) - H_{12}(q,\delta)K\delta + H_{11}(q,\delta)u \\
\ddot{\delta} &= -H_{21}(q,\delta)(f_1(q,\dot{q}) + g_1(q,\dot{q},\delta,\dot{\delta})) \\
&\quad - H_{22}(q,\delta)(f_2(q,\dot{q}) + g_2(q,\dot{q},\delta,\dot{\delta})) - H_{22}(q,\delta)K\delta + H_{21}(q,\delta)u
\end{aligned}$$

$$(2.2)$$

where $q \in \mathbf{R}^n$ is the vector of joint position variables, $\delta \in \mathbf{R}^m$ is the vector of flexible modes, f_1, f_2, g_1, and g_2 are the terms due to gravity, Coriolis, and centripetal forces, H is the inverse of the positive–definite mass matrix M such that $M_{i,j}, H_{i,j}, \ i,j = 1,2$ are the submatrices corresponding to the q and δ vectors, and K is the positive definite stiffness matrix. Let us define the new state variables

$$x_1 = q, \qquad x_2 = \dot{q}, \qquad z_1 = \frac{\delta}{\varepsilon^2}, \qquad z_2 = \frac{\dot{\delta}}{\varepsilon} \qquad (2.3)$$

where ε is the singular perturbation parameter defined [43] as

$$\varepsilon^2 = \frac{1}{\lambda_{min}(H_{220}K)}. \qquad (2.4)$$

In (2.4), $\lambda_{min}(H_{220}K)$ is the lower bound of the minimum eigenvalue of the H_{22} submatrix evaluated at $\delta = 0$ (i.e. $H_{220} = H_{22}(q,0)$) over the range in which q varies. The system described by (2.2) may then be written as

$$
\begin{aligned}
\dot{x}_1 &= x_2 \\
\dot{x}_2 &= a(x_1, x_2, \varepsilon^2 z_1, \varepsilon z_2) - A(x_1, \varepsilon^2 z_1)z_1 + H_{11}(x_1, \varepsilon^2 z_1)u \quad (2.5) \\
\varepsilon \dot{z}_1 &= z_2 \\
\varepsilon \dot{z}_2 &= b(x_1, x_2, \varepsilon^2 z_1, \varepsilon z_2) - B(x_1, \varepsilon^2 z_1)z_1 + H_{21}(x_1, \varepsilon^2 z_1)u \quad (2.6)
\end{aligned}
$$

where $x_1, x_2 \in \mathbf{R}^n$, $z_1, z_2 \in \mathbf{R}^m$ and

$$
\begin{aligned}
a(x_1, x_2, \varepsilon^2 z_1, \varepsilon z_2) &= -H_{11}f_1 - H_{12}f_2 - H_{11}g_1 - H_{12}g_2 \\
b(x_1, x_2, \varepsilon^2 z_1, \varepsilon z_2) &= -H_{21}f_1 - H_{22}f_2 - H_{21}g_1 - H_{22}g_2 \\
A(x_1, \varepsilon^2 z_1) &= \frac{H_{12}(x_1, \varepsilon^2 z_1)K}{\lambda_{min}(H_{220}K)} \\
B(x_1, \varepsilon^2 z_1) &= \frac{H_{22}(x_1, \varepsilon^2 z_1)K}{\lambda_{min}(H_{220}K)}. \qquad (2.7)
\end{aligned}
$$

Defining the tip positions as outputs, the output vector y is written as [79]

$$y = x_1 + \Psi \varepsilon^2 z_1 \qquad (2.8)$$

where Ψ is an $n \times n$ matrix depending on the shape function used in the original model given by (2.2) (see Appendix B for further details). In this analysis it is assumed that the vibrations are in the lateral plane of each joint axis. Let $x(t, \varepsilon)$ and $z(t, \varepsilon)$ denote the solutions of (2.5)–(2.6). In the $(2n + 2m)$–dimensional state space of (2.5)–(2.6), a $2n$–dimensional manifold $\mathcal{M}_\varepsilon$, depending on the scalar ε, defined by

$$\mathcal{M}_\varepsilon : \quad z = h(x, u, \varepsilon) \qquad (2.9)$$

is said to be an integral manifold of (2.5)–(2.6) if given

$$z(t_0, \varepsilon) = h(x(t_0, \varepsilon), u, \varepsilon)$$

it then follows that $z(t, \varepsilon) = h(x(t, \varepsilon), u, \varepsilon)$ for all $t \geq t_0$, where $z^T = [z_1^T \ z_2^T]$. Substituting h from (2.9) in (2.6) leads to a partial differential equation for h that is referred to in the literature as the *manifold condition* [45], [81], [53]. However, an approximate solution may be found by a series expansion of u and h in terms of ε. The ε^2 term in (2.8) suggests that the expansions of $h(x, u, \varepsilon)$ and u are required at least up to ε^3 terms if output feedback is to be used for control. This will become obvious shortly from equation (2.23). Let us now express the control input u according to

$$u = u_s(x, \varepsilon, t) + u_f(x, z) \tag{2.10}$$

where

$$u_s = u_0 + \varepsilon u_1 + \varepsilon^2 u_2 + O(\varepsilon^3) \tag{2.11}$$

with u_0, u_1, and u_2 to be designed subsequently. It is further assumed that u_f is zero on the second order corrected slow manifold, that is, up to $O(\varepsilon^3)$. Expanding z_1 and z_2 in (2.9) as

$$z_1 := h_1(x, u, \varepsilon) = h_{10} + \varepsilon h_{11} + \varepsilon^2 h_{12} + O(\varepsilon^3)$$
$$z_2 := h_2(x, u, \varepsilon) = h_{20} + \varepsilon h_{21} + \varepsilon^2 h_{22} + O(\varepsilon^3) \tag{2.12}$$

and substituting (2.12) in (2.6) gives

$$
\begin{aligned}
&\varepsilon \left(\dot{h}_{10} + \varepsilon \dot{h}_{11} + \varepsilon^2 \dot{h}_{12} + O(\varepsilon^3) \right) = h_{20} + \varepsilon h_{21} + \varepsilon^2 h_{22} + O(\varepsilon^3) \\
&\varepsilon \left(\dot{h}_{20} + \varepsilon \dot{h}_{21} + \varepsilon^2 \dot{h}_{22} + O(\varepsilon^3) \right) = \\
&\quad b \left(x_1, x_2, \varepsilon^2 (h_{10} + O(\varepsilon)), \varepsilon(h_{20} + \varepsilon h_{21} + O(\varepsilon^2)) \right) \\
&\quad - \ B(x_1, \varepsilon^2 (h_{10} + O(\varepsilon))(h_{10} + \varepsilon h_{21} + O(\varepsilon^2)) \\
&\quad + \ H_{21}(x_1, \varepsilon^2 (h_{10} + O(\varepsilon)) u.
\end{aligned}
\tag{2.13}
$$

Equating the terms with the same powers in ε on both sides of the above equations up to ε^3 and using (2.11) gives

$$
\begin{aligned}
h_{20} &= 0 \\
h_{21} &= \dot{h}_{10} \\
h_{22} &= \dot{h}_{11} \\
h_{10} &= B^{-1}(x_1, 0)(b(x_1, 0, 0) + H_{21}(x_1, 0)u_0)
\end{aligned}
$$

$$h_{11} = B^{-1}(x_1,0)H_{21}(x_1,0)u_1$$

$$h_{12} = B^{-1}(x_1,0)(-\dot{h}_{21} + \frac{\partial b}{\partial \delta}\mid_{\delta,\dot{\delta}=0} h_{10} + \frac{\partial b}{\partial \dot{\delta}}\mid_{\delta,\dot{\delta}=0} h_{21}$$

$$- (\sum_{i=1}^{m} \frac{\partial B}{\partial \delta_i}\mid_{\delta=0} h_{10i})h_{10} + H_{21}(x_1,0)u_2$$

$$+ (\sum_{i=1}^{m} \frac{\partial H_{21}}{\partial \delta_i}\mid_{\delta=0} h_{10i})u_0) \tag{2.14}$$

where h_{10i} is the i-th element of h_{10}. Substituting z_1 and z_2 from (2.12) into (2.5) yields the dynamics of the *exact* slow subsystem restricted to the $\mathcal{M}_\varepsilon$ manifold [43], [90], [45], [46], [53]. Keeping the $O(\varepsilon^2)$ terms and assuming that the fast control u_f is inactive on the $O(\varepsilon^2)$ approximate manifold, or more precisely, the second order manifold represented by $\mathcal{M}_2$: $z = h^0(x,u_0,0) + \varepsilon h^1(x,u_0,u_1,\varepsilon) + O(\varepsilon^2)$, the *second order corrected slow* subsystem is obtained as

$$\dot{x}_1 = x_2$$

$$\dot{x}_2 = M_{110}^{-1}(u_0 + \varepsilon u_1 + \varepsilon^2 u_2 - f_1(x_1,x_2))$$

$$- \varepsilon^2 d(x_1,x_2,\dot{x}_2\mid_{\varepsilon=0},u_0,\dot{u}_0) \tag{2.15}$$

where $M_{110} = M_{11}(q,0)$ [1], and $d(.)$ is given by

$$d(x_1,x_2,\dot{x}_2\mid_{\varepsilon=0},u_0,\dot{u}_0) = H_{110}(G_{12}h_{10} + G_{11}h_{21})$$

$$+ (\sum_{i=1}^{m} \frac{\partial H_{11}}{\partial \delta_i}\mid_{\delta=0} h_{10i})f_1 + (\sum_{i=1}^{m} \frac{\partial H_{12}}{\partial \delta_i}\mid_{\delta=0} h_{10i})f_2$$

$$+ H_{120}(G_{22}h_{10} + G_{21}h_{21}) + (\sum_{i=1}^{m} \frac{\partial A}{\partial \delta_i}\mid_{\delta=0} h_{10i})h_{10}$$

$$- (\sum_{i=1}^{m} \frac{\partial H_{11}}{\partial \delta_i}\mid_{\delta=0} h_{10i})u_0 + H_{120}H_{220}\{-\dot{h}_{21} + \frac{\partial b}{\partial \delta}\mid_{\delta,\dot{\delta}=0} h_{10}$$

$$+ \frac{\partial b}{\partial \dot{\delta}}\mid_{\delta,\dot{\delta}=0} h_{21} - (\sum_{i=1}^{m} \frac{\partial B}{\partial \delta_i}\mid_{\delta=0} h_{10i})h_{10}$$

$$+ ((\sum_{i=1}^{m} \frac{\partial H_{21}}{\partial \delta_i}\mid_{\delta=0} h_{10i})u_0\} \tag{2.16}$$

---

[1] In general $M_{ij0} = M_{ij}(q,0)$ and $H_{ij0} = H_{ij}(q,0)$ for $i = 1,2$, $j = 1,2$.

with the G_{ij}, $i, j = 1, 2$ matrices in (2.16) defined as

$$G_{11}(x_1, x_2) = \frac{\partial g_1}{\partial \delta} |_{\delta, \dot{\delta}=0} \qquad G_{21}(x_1, x_2) = \frac{\partial g_2}{\partial \delta} |_{\delta, \dot{\delta}=0}$$

$$G_{12}(x_1, x_2) = \frac{\partial g_1}{\partial \dot{\delta}} |_{\delta, \dot{\delta}=0} \qquad G_{22}(x_1, x_2) = \frac{\partial g_2}{\partial \dot{\delta}} |_{\delta, \dot{\delta}=0} . \qquad (2.17)$$

It should be pointed out that in the calculation of $\dot{h}_{10}$ and $\ddot{h}_{10}$, whenever $\dot{x}_1$ and $\dot{x}_2$ are required, they are obtained from the controlled *rigid* model (i.e. at $\varepsilon \equiv 0$) given by

$$\dot{x}_1 = x_2$$
$$\dot{x}_2 = a(x_1, x_2, 0, 0) - A(x_1, 0)h_{10} + H_{11}(x_1, 0)u_0. \qquad (2.18)$$

The exact fast variable z will deviate from the second order manifold $\mathcal{M}_2$. Representing this deviation by $\tilde{z}_1$, $\tilde{z}_2$ according to

$$\tilde{z}_1 = z_1 - (h_{10} + \varepsilon h_{11} + \varepsilon^2 h_{12})$$
$$\tilde{z}_2 = z_2 - (h_{20} + \varepsilon h_{21} + \varepsilon^2 h_{22}) \qquad (2.19)$$

and substituting (2.19) in (2.6) results in the *exact fast* subsystem described by

$$\varepsilon \dot{\tilde{z}}_1 = \tilde{z}_2 - \varepsilon^3 \dot{h}_{12}$$
$$\varepsilon \dot{\tilde{z}}_2 = -(B(x_1, 0) + \varepsilon^2 \sum_{i=1}^{m} \frac{\partial B}{\partial \delta_i} |_{\delta=0} h_{10i})\tilde{z}_1$$
$$+ (H_{210} + \varepsilon^2 \sum_{i=1}^{m} \frac{\partial H_{21}}{\partial \delta_i} |_{\delta=0} h_{10i})u_f + O(\varepsilon^3). \qquad (2.20)$$

By neglecting the $O(\varepsilon^3)$ terms, the *corrected fast* subsystem is now governed by

$$\varepsilon \dot{\tilde{z}}_1 = \tilde{z}_2$$
$$\varepsilon \dot{\tilde{z}}_2 = -(B(x_1, 0) + \varepsilon^2 \sum_{i=1}^{m} \frac{\partial B}{\partial \delta_i} |_{\delta=0} h_{10i})\tilde{z}_1$$
$$+ (H_{210} + \varepsilon^2 \sum_{i=1}^{m} \frac{\partial H_{21}}{\partial \delta_i} |_{\delta=0} h_{10i})u_f. \qquad (2.21)$$

Our aim is to use output feedback to stabilize the fast subsystem (2.21) so that the second order manifold is an attractive set. To this end, let us consider y from (2.8) when z_1 is defined from (2.19). This defines the actual output *restricted* to $O(\varepsilon^3)$ manifold by

$$y_{res} = x_1 + \varepsilon^2 \Psi(\tilde{z}_1 + h_{10}). \qquad (2.22)$$

Thus, we may now define the *slow* and *fast* outputs y_s and y_f, respectively according to

$$
\begin{aligned}
y_s : &= x_1 + \varepsilon^2 \Psi h_{10} \\
y_f : &= \varepsilon^2 \Psi \tilde{z}_1
\end{aligned}
\qquad (2.23)
$$

so that $y_{res} = y_s + y_f$. Note that y_s can be obtained from measurements of the joint variables (positions and velocities). If the total tip deflection y is measured, then y_f can be constructed formally from $y_f = y - y_s$. Thus, if $y - y_s$ is used to obtain y_f, there will always be an $O(\varepsilon^3)$ error term due to the neglected unmodeled terms.

2.3. Slow and Fast Subsystem Control Strategies

In this section, we will develop control strategies based on output feedback for the corrected slow and fast subsystems described by (2.15) and (2.21), respectively. The outputs of the subsystems are given by (2.23).

Slow Subsystem Control Strategy

Consider the input–output representation of the second order corrected slow subsystem described by (2.15) with output given by y_s in (2.23), i.e.,

$$
\begin{aligned}
\ddot{y}_s = \ & M_{110}^{-1}(u_0 + \varepsilon u_1 + \varepsilon^2 u_2 - f_1(x_1, x_2)) \\
& - \varepsilon^2 d(x_1, x_2, \dot{x}_2 \mid_{\varepsilon=0}, u_0, \dot{u}_0) + \varepsilon^2 \Psi \ddot{h}_{10}
\end{aligned}
\qquad (2.24)
$$

The objective, as proposed in [43], [30], [90] is to design the control terms u_0, u_1 and u_2 such that the resulting closed–loop system has asymptotically stable zero dynamics. Towards this end, taking

$$
\begin{aligned}
u_0 &= M_{110}v_0 + f_1(x_1, x_2) \\
u_1 &= M_{110}v_1 \\
u_2 &= M_{110}(v_2 + d(x_1, x_2, \dot{x}_2 \mid_{\varepsilon=0}, u_0, \dot{u}_0) - \Psi \ddot{h}_{10}). \quad (2.25)
\end{aligned}
$$

will render (2.24) into

$$\ddot{y}_s = v_0 + \varepsilon v_1 + \varepsilon^2 v_2 \tag{2.26}$$

where v_0, v_1, and v_2 are new inputs to be defined subsequently. Note that with the above choice of control laws, h_{10}, h_{11}, and h_{12} are now expressed in terms of v_0, v_1, and v_2, respectively. Now let us choose

$$\begin{aligned} v_1 &= A_1 \dot{v}_0 \\ v_2 &= A_2 \ddot{v}_0 \end{aligned} \tag{2.27}$$

which when substituted in (2.26) yield

$$\ddot{y}_s = v_0 + \varepsilon A_1 \dot{v}_0 + \varepsilon^2 A_2 \ddot{v}_0 := v_s. \tag{2.28}$$

This is the new representation of the corrected slow subsystem with its zero dynamics given by

$$\dot{v} = A_v v \tag{2.29}$$

where $v^T = [v_0^T \; v_1^T]$, and

$$A_v = \begin{bmatrix} 0 & A_1^{-1} \\ -\frac{A_1 A_2^{-1}}{\varepsilon^2} & -\frac{A_1 A_2^{-1}}{\varepsilon} \end{bmatrix}. \tag{2.30}$$

The matrices A_1 and A_2 are chosen such that the zero–dynamics are asymptotically stable. This can always be guaranteed, for instance by taking $A_1 = I$ and A_2 any positive definite matrix. The tracking objective for system (2.28) is now stated as follows: Design the control law v_s such that the resulting closed–loop output y_s and its higher order derivatives follow prescribed desired trajectories.

Let v_s in (2.28) be defined as

$$v_s = \ddot{y}_r - K_d(\dot{y}_s - \dot{y}_r) - K_p(y_s - y_r) \tag{2.31}$$

where y_r, $\dot{y}_r$, and $\ddot{y}_r$ define the reference trajectory to be tracked, and further define the tracking error signals by

$$e_1 = y_s - y_r, \quad e_2 = \dot{y}_s - \dot{y}_r, \quad e^T = [e_1^T \; e_2^T]. \tag{2.32}$$

The error dynamics are now given by

$$\dot{e} = A_e e \tag{2.33}$$

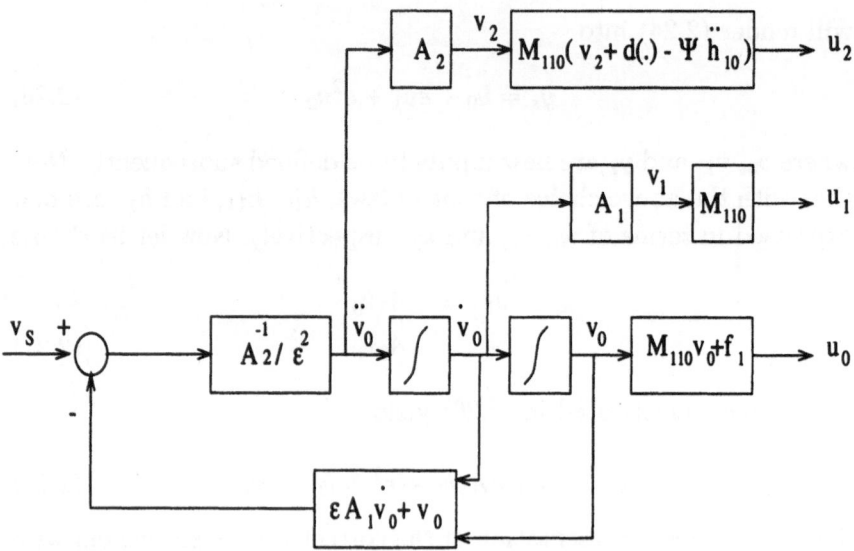

Figure 2.1. Block diagram of the *corrected slow* subsystem control strategy.

where

$$A_e = \begin{bmatrix} 0 & I \\ -K_p & -K_d \end{bmatrix}. \qquad (2.34)$$

Substituting (2.31) into (2.28) and using (2.30) and (2.32) yields

$$\dot{v} = A_v v + b_v(\ddot{y}_r, e, \varepsilon) \qquad (2.35)$$

where

$$b_v(\ddot{y}_r, e, \varepsilon) = \begin{bmatrix} 0 \\ \frac{A_1 A_2^{-1}}{\varepsilon^2}(\ddot{y}_r - K_d e_2 - K_p e_1) \end{bmatrix}. \qquad (2.36)$$

A block diagram of the *corrected slow* subsystem control strategy is shown in Figure 2.1.

Fast Subsystem Control Strategy

Consider the corrected fast dynamics (2.21) with the fast output y_f defined by (2.23). We are interested in using dynamic output feedback to stabilize the fast dynamics. To this end, (2.21) is written in the form

$$\varepsilon \dot{\tilde{z}} = A_f(x_1, x_2, \varepsilon^2, v_0)\tilde{z} + B_f(x_1, x_2, \varepsilon^2, v_0)u_f \qquad (2.37)$$

where $\tilde{z}^T = [\tilde{z}_1^T \quad \tilde{z}_2^T]$ and

$$
\begin{aligned}
A_f(\cdot) &= A_{f0} + \varepsilon^2 A_{f1}(x_1, x_2, v_0) \\
&:= \begin{bmatrix} 0 & I \\ -B(x_1, 0) & 0 \end{bmatrix} + \varepsilon^2 \begin{bmatrix} 0 & 0 \\ -\sum_{i=1}^{m} \frac{\partial B}{\partial \delta_i}\big|_{\delta=0} h_{10i} & 0 \end{bmatrix} \\
B_f(\cdot) &= B_{f0} + \varepsilon^2 B_{f1}(x_1, x_2, v_0) \\
&:= \begin{bmatrix} 0 \\ -H_{210} \end{bmatrix} + \varepsilon^2 \begin{bmatrix} 0 \\ \sum_{i=1}^{m} \frac{\partial H_{21}}{\partial \delta_i}\big|_{\delta=0} h_{10i} \end{bmatrix}.
\end{aligned}
\tag{2.38}
$$

The output y_f may also be written as

$$
y_f = C_f \tilde{z} \tag{2.39}
$$

where the $n \times 2m$ output matrix C_f is given by

$$
C_f = [\varepsilon^2 \Psi \quad 0]. \tag{2.40}
$$

Since the slow variables are treated as *frozen parameters* [81] in (2.37), the above system is *linear* in terms of $\tilde{z}$. Considering the $O(1)$ terms in $A_f(\cdot)$ and $B_f(\cdot)$ in (2.38), the general configuration for the dynamic output feedback controller is proposed as

$$
\varepsilon \dot{w} = F(x_1)w + G(x_1)y_f \tag{2.41}
$$
$$
u_f = M(x_1)w + N(x_1)y_f \tag{2.42}
$$

where $w \in \mathbf{R}^l$, and matrices $F(x_1)$ $(l \times l)$, $G(x_1)$ $(l \times n)$, $M(x_1)$ $(n \times l)$, and $N(x_1)$ $(n \times n)$ are to be selected so that the resulting closed–loop corrected fast subsystem is asymptotically stable. In order to have a controller which is robust to higher order unmodeled dynamics and measurement noise, there should be no feed–through of the output in the control law (i.e., $N(x_1) \equiv 0$) [47]. Thus, augmenting the fast dynamics (2.37) with (2.41) and using the control law (2.42) yields the closed–loop corrected fast subsystem

$$
\varepsilon \dot{\eta} = (A_\eta(x_1) + \varepsilon^2 A_{\eta\varepsilon^2}(x_1, x_2, v_0))\eta \tag{2.43}
$$

where

$$
\begin{aligned}
\eta^T &= [\tilde{z}^T \quad w^T] \\
A_\eta(x_1) &= \begin{bmatrix} A_{f0}(x_1) & B_{f0}(x_1)M(x_1) \\ G(x_1) & F(x_1) \end{bmatrix} \\
A_{\eta\varepsilon^2}(x_1, x_2, v_0) &= \begin{bmatrix} A_{f1}(x_1, x_2, v_0) & B_{f1}(x_1, x_2, v_0)M(x_1) \\ 0 & 0 \end{bmatrix}
\end{aligned}
$$

$$
\tag{2.44}
$$

and which by design is guaranteed to be asymptotically stable.

A stability analysis for the full–order system is performed by considering the open–loop system (2.5)–(2.6) and the control laws (2.10), (2.11), (2.25), (2.35), (2.41), and (2.42), and by relaxing the *frozen parameter* assumption in treating the corrected fast subsystem. This analysis is now summarized in the following theorem:

Theorem 2.1 *Let the control laws (2.10), (2.11), (2.25), (2.35), (2.41), and (2.42) be applied to the open–loop nonlinear system (2.5)-(2.6). Assuming that the desired reference trajectories and their time derivatives (at least up to order 2) are continuous and bounded, it then follows that the trajectories of e, η, and $\varepsilon^2 v$ converge to a residual set of order $O(\varepsilon^2)$ if the perturbation parameter ε belongs to the interval $(0, \varepsilon_{max})$ with ε_{max} obtained from matrix Λ in (A.11), and further, provided that certain norm conditions on the vectors b_e, b_v, b_η, the matrices A_e, A_v, A_η, $F(x_1)$, $G(x_1)$, and $M(x_1)$ defined by (A.7)–(A.9) are satisfied.*

Proof: The above result is proved in Appendix A.1 by utilizing a Lyapunov stability analysis.

Remark 1

Using the above theorem, it can also be shown that the tip position and velocity tracking errors are $O(\varepsilon^2)$. To show this, consider for example $y - y_r$. From (2.8) and (2.12), it follows that $y = x_1 + \Psi \varepsilon^2 h_{10} + O(\varepsilon^3)$, which from (2.23) may be written as $y = y_s + O(\varepsilon^3)$. Consequently, as shown in the theorem, since $y_s \rightarrow y_r + O(\varepsilon^2)$, it then follows that $y \rightarrow y_r + O(\varepsilon^2)$. A similar result also holds for $\dot{y}$, that is, $\dot{y} \rightarrow \dot{y}_r + O(\varepsilon^2)$.

Remark 2 (Robustness considerations)

The significance of the stability analysis presented in Appendix A.1 is that it can provide the designer with guidelines for selecting controller gain matrices for a more robust design. Towards this end, consider the elements of the matrix Λ that are affected by the control gain matrices K_p, K_d, A_1, A_2, $M(x_1)$, $F(x_1)$, and $G(x_1)$. In general, to ensure better robustness, the off–diagonal terms in Λ should be decreased and the diagonal terms should be increased. A closer inspection of Λ reveals that decreasing $\gamma_{pd}\|P_v\|$, $\gamma_E\|P_e\|$, and $2\gamma_\eta\|P_\eta\| + l_3$ fulfills the aforementioned goal. It can also be concluded from these terms that choosing the gain matrices to reduce the norms $\|P_\eta\|$, $\|\partial P_\eta(x_1)/\partial x_1\|$, $\|A_1 A_2^{-1}[K_p \; K_d]\|$, $\|P_v\|$, $\|M(x_1)\|$, and $\|P_e\|$ will generally result in a more robust closed–loop system. Of these terms, the matrix $M(x_1)$ was found experimentally to have a significant effect on robustness. This may be attributed to the fact that it affects the γ_E term (see (A.8)) which in turn appears as an $O(1)$ off–diagonal term in Λ.

Numerical Simulations: A Two–link Flexible Manipulator

A two–link planar manipulator is considered in which the first link is rigid and the second link is flexible. The main reasons for investigating this system are that it contains strong nonlinear coupling terms in addition to being non–minimum phase. The two–link data are as follows

$$l_1 = 0.2, \; l_2 = 0.6m, \; a_1 = 1.3cm \times 3.0, \; a_2 = 0.88mm \times 5.0cm$$
$$\rho = 7980kg/m^3 \; (Steel), \; M_1 = 1kg, \; M_2 = 0.25kg, \; \varepsilon = 0.03$$
$$E = 190 \times 10^9 N/m^2, J_1 = J_2 = 0.002, J_h = 3 \times 10^{-5} kgm^2$$

where l_1 (l_2), a_1 (a_2), M_1 (M_2), J_h (J_1, J_2), ρ and E denote link lengths, cross sectional areas, masses at the end points of each link, mass moments of inertia (hub, second joint, load), mass density, and Young's modulus of elasticity, respectively.

The first two flexible modes of this system when linearized around zero joint angles are 5.6 and 27.6 Hz. The roots corresponding to the linearized zero dynamics (when the tip position is taken as the output) are at $\pm j76.9$ and ± 16.0. The slow control components of u_2 (e.g. $d(\cdot)$) were obtained by $MAPLE$. The matrices $F(x_1)$, $G(x_1)$, and $M(x_1)$ are obtained to place the poles of $A_\eta(x_1)$ in the left–half of the complex plane. This is achieved by obtaining the two gain matrices $K(x_1)$ and $L(x_1)$ as follows: The matrix $K(x_1)$ is obtained by solving a pole placement problem for the pair $(A_{f0}(x_1), B_{f0}(x_1))$ with a prescribed set of pole locations P_{K_d} selected to be in the left–half of the complex plane. Similarly, the matrix $L(x_1)$ is obtained by placing the poles of the pair $(A_{f0}^T(x_1), C_f^T)$ to be at a prescribed set of pole locations P_{L_d} selected to be in the left–half of the complex plane. This is equivalent to solving the stabilization problem of the triple $(A_{f0}(x_1), B_{f0}(x_1), C_f)$ using a Luenberger observer.

The stabilization of $A_\eta(x_1)$ is achieved by dividing the workspace trajectory into ten segments and solving a dynamic pole placement problem for each segment such that the eigenvalues of $A_\eta(x_1)$ are

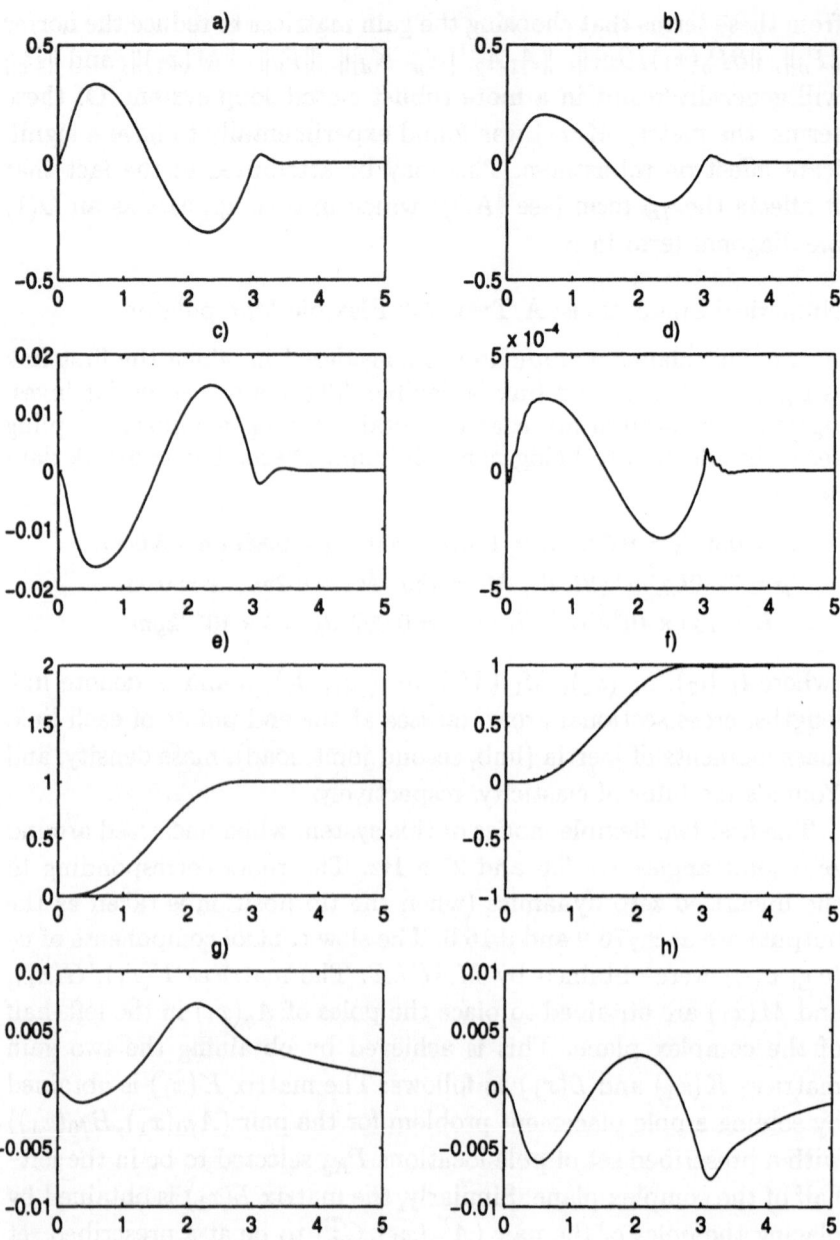

Figure 2.2. Simulation results for a flexible two–link manipulator using the proposed corrected controller (Horizontal axis: time (s)):   a) First joint torque input (Nm)   b) Second joint torque input (Nm)   c) First flexible mode (m) d) Second flexible mode (m)   e) First joint angle (—) and reference trajectory (···) (rad)   f) Tip position of second link (—) and reference trajectory (···) (rad)   g) Tracking error of the first link joint position, $y_{1r} - y_1$ (rad)   h) Tip position tracking error of the second link, $y_{2r} - y_2$ (rad).

at prescribed locations in the left–half of the complex plane. The gain matrices $F(x_1)$, $G(x_1)$, and $M(x_1)$ were then obtained by linear interpolation. In this way, the maximum real part of the eigenvalues of $A_\eta(x_1)$ is ensured to be negative when x_1 (second joint position variable) lies in the region of interest. It should be pointed out that the use of state dependent gain matrices $F(x_1)$, $G(x_1)$, and $M(x_1)$ does not, in general, guarantee stability of the closed–loop system and further interconnection conditions (see (A.7)–(A.9) in Appendix A.1) should be satisfied.

A qualitative measure that is observed from the Lyapunov stability analysis (cf. Remark 2 in Section 2.3) is utilized to improve the robustness of the closed–loop system. Specifically, lowering the norm of the matrix $M(x_1)$ was found to have a significant effect on the stability condition. Thus, to decrease this norm, the first row of this matrix was set to zero (which is equivalent to de–activating the fast control term for the first–link actuator). This is justified by noting that the first actuator has a lower accessibility to the vibrational modes of the second link and is likely to destabilize the closed–loop system because of higher gain requirements. The roots of the linearized zero–dynamics are located at ± 22.51 and ± 64.70, and the other gain matrices are: $A_1 = A_2 = 5I_{2\times 2}$, $K_p = I_{2\times 2}$, $K_d = 2I_{2\times 2}$.

The simulation results for quintic reference trajectories are shown in Figure 2.2. It is observed that the tracking errors $y_{1r} - y_1$ and $y_{2r} - y_2$ are both of order $O(\varepsilon^2)$ as expected (recall that $\varepsilon = 0.03$). The magnitude of the fast control u_f is relatively small compared to the slow control u_s, that is of the same order of magnitude as the control required for rigid–body motion. The maximum deflection of the arm is plotted in Figure 2.3 together with the tip deflection error at each instant of time.

A comparison of the performance of the proposed controller with that of other methods in the literature can be made by choosing the composite controller to consist of a *slow* control law that is designed based on the *rigid body* model (e.g. [8], [49]) and a fast control law that is identical to the one used in the proposed controller. Specifically, the slow controller u_0 in (2.25) is obtained by setting $\varepsilon = 0$ in (2.10), (2.11), and in y_s given by (2.23), and in $\ddot{y}_s$ given by (2.28) to yield

$$u_0 = M_{110}(\ddot{y}_r - K_d(x_2 - \dot{y}_r) - K_p(x_1 - y_r)) + f_1(x_1, x_2).$$

Simulation results for the same K_p and K_d as before are shown in

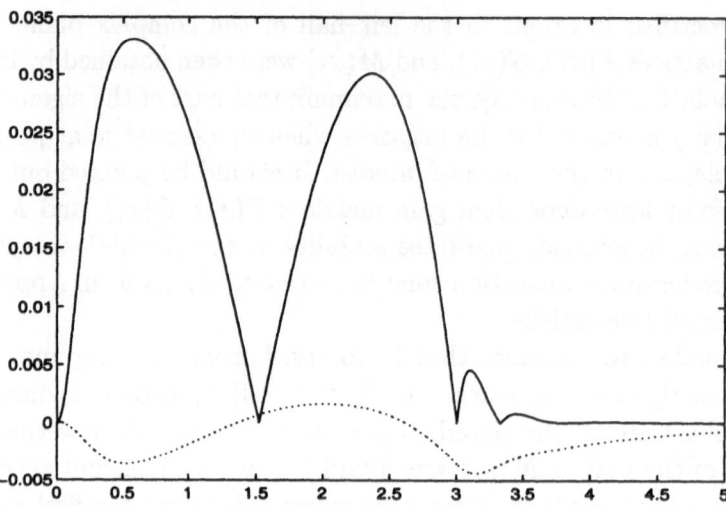

Figure 2.3. Simulation results for the proposed corrected control scheme: Maximum absolute value of the arm deflection (—), in meters, at each instant computed among twenty equidistant points on the second link, and tip deflection tracking error, $l_2(y_{2r} - y_2)$ ($\cdots$) (m).

Figure 2.4. As can be seen, this composite controller results in worse tracking error performance as compared to the proposed controller. The maximum deflection of the arm is plotted in Figure 2.5 together with the tip deflection error at each instant of time. The two case studies show that the proposed controller has been successful in providing a stable control action in addition to smaller tracking errors (the maximum absolute error is 7.3 times smaller as seen from Figures 2.2h and 2.4h). A quantitative measure to evaluate the tracking performance of the controllers can be defined as the ratio of maximum tip deflection error to the maximum arm deflection during the whole trajectory. For the proposed control scheme, this ratio is 0.13 (Figure 2.3), while for the rigid–body based *slow* control, a ratio of 0.86 is obtained (Figure 2.5). In other words, the proposed control scheme results in an improvement of 6.6 times (0.86/0.13) in the above ratio.

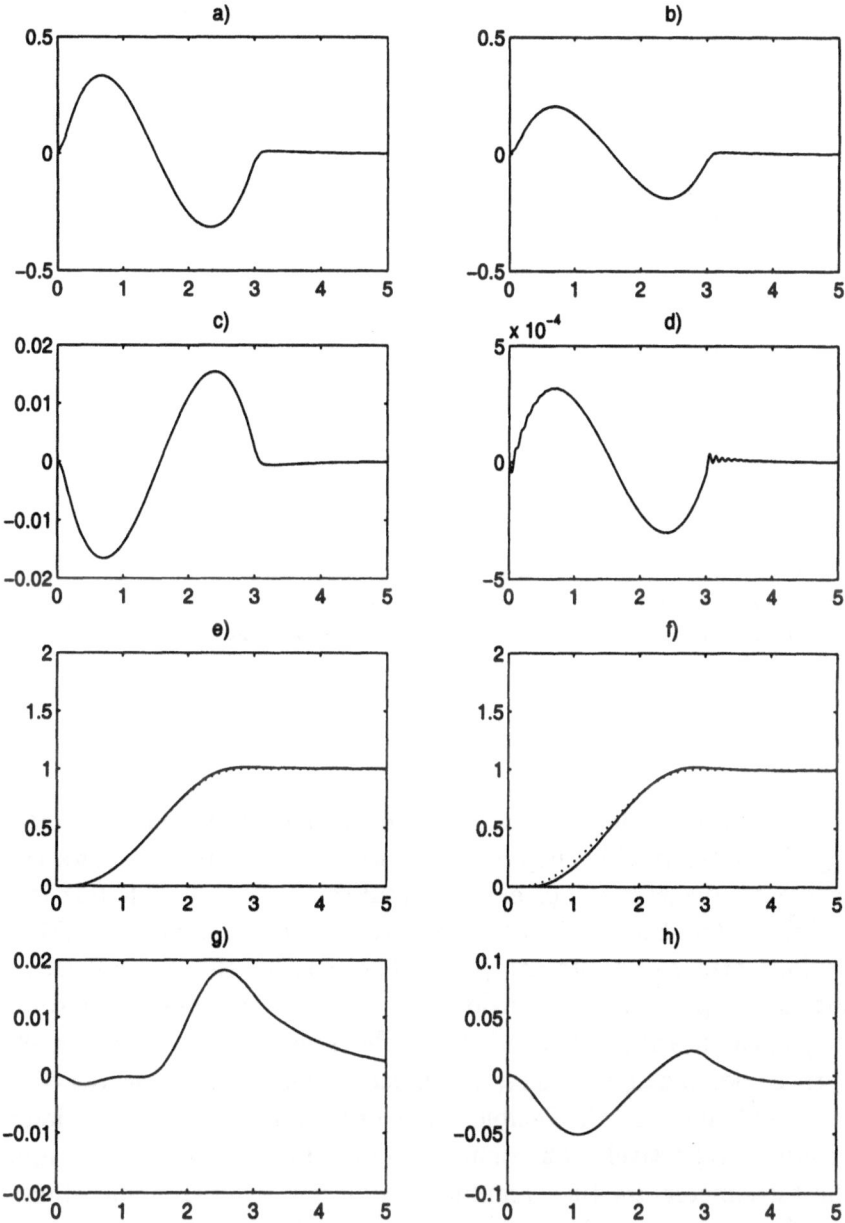

Figure 2.4. Simulation results for *slow* control that is based on rigid–body model plus fast control (uncorrected scheme)(Horizontal axis: time (s)):   a) First joint torque input (Nm)   b) Second joint torque input (Nm)   c) First flexible mode (m)   d) Second flexible mode (m)   e) First joint angle (—) and reference trajectory ($\cdots$) (rad)   f) Tip position of second link (—) and reference trajectory ($\cdots$) (rad)   g) Tracking error of the first link joint position, $y_{1r} - y_1$ (rad)   h) Tip position tracking error of the second link, $y_{2r} - y_2$ (rad).

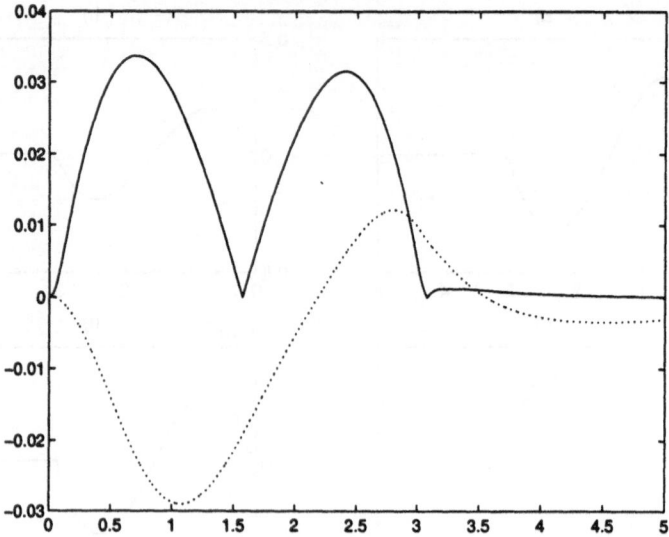

Figure 2.5. Simulation results for the scheme based on rigid–body slow control plus the fast control (uncorrected scheme): Maximum absolute value of the arm deflection (—) at each instant computed among twenty equidistant points on the second link, and tip deflection tracking error, $l_2(y_{2r} - y_2)$ ($\cdots$) (m).

2.4. Experimental Results

In this section, the practical implementation of the control strategy discussed in this chapter is investigated. Figure 2.6 shows the schematic diagram of our experimental setup. The flexible link is a stainless–steel $60cm \times 5cm \times 0.9mm$ rectangular bar with a $0.251kg$ payload attached to its end point. The mass of the bar is $0.216kg$ that is comparable to its payload. The first three measured natural frequencies are at 5.5, 20, and 45 Hz. The sensory equipment consists of three strain gauge bridges, a tachometer, and a shaft–encoder that are used to measure the flexible modes of the link, joint rate, and joint position, respectively. The signals from the strain gauge bridges and the tachometer are then amplified using low–drift amplifier stages and further passed through anti–aliasing filters. These signals are then fed into the *XVME-500/3* analog input module from *Xycom*. The actuator is a *5113 Pittman* DC brushless servomotor which is driven by a *503 Copley* PWM servo–drive amplifier.

The digital hardware has been selected based on the idea of a *re-configurable* sensor–based control application. The *Chimera 3.2 Real Time Operating System* [73] is used as a local operating system in

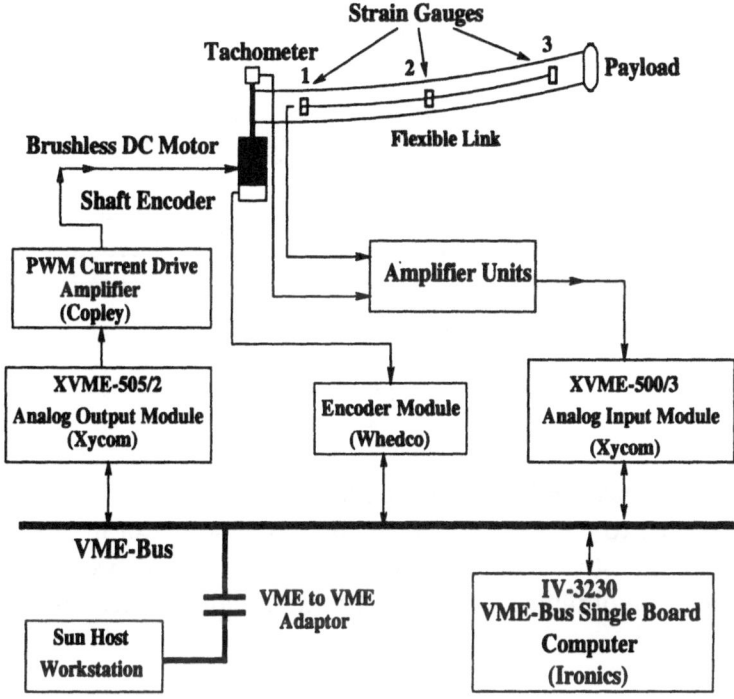

Figure 2.6. Experimental setup for the flexible arm.

conjunction with a global *UNIX* environment. It can execute on one
or more single board computers in a VMEbus–based system. In Figure 2.6, the *Chimera 3.1* kernel is running on the Ironics MC68030
processor with $33MHz$ clock frequency and a floating–point co-
processor. This processor, along with the analog and digital output
modules, are used for data acquisition as well as computation of the
control algorithm. The code to run under *Chimera 3.1* is written in
C.

The tip deflection is constructed based on the measurements obtained from the strain gauges. Considering a point x along the link,
its deflection as a function of time can be written as

$$w(x, t) = \sum_{i=1}^{m} \phi_i(x)\delta_i(t) \qquad (2.45)$$

where m is the number of flexible modes, $\phi_i(x)$ is the i-th mode
shape function, and $\delta_i(t)$ is the i-th flexible mode. The longitudinal

strain at point x is then obtained from

$$\epsilon(x,t) = \frac{D}{2} \frac{\frac{\partial^2 w}{\partial x^2}}{\sqrt{1 + (\frac{\partial w}{\partial x})^2}} \qquad (2.46)$$

where D is the thickness of the beam at point x and the remaining term is the reciprocal of the radius of curvature at x. An approximation to the above formula can be obtained by noting that for typical motions of the link $\partial w/\partial x$, is small. Thus

$$\epsilon(x,t) \approx \frac{D}{2} \sum_{i=1}^{m} \frac{\partial^2 \phi(x)}{\partial x^2} \delta_i(t). \qquad (2.47)$$

In order to ensure a better approximation, three points on the link were selected to yield a small norm for the denominator in (2.46). By measuring the strains at the three points on the link, two deflection modes can be obtained using the Moore pseudo–inverse formula as shown in [64].

2.4.1. Model Validation

In order to evaluate the accuracy of the open–loop dynamic model of the flexible–link system, frequency responses of the link were obtained from the strain gauge outputs to the command torque input for a sufficiently small input as long as the output signals remain undistorted. The nonlinear dynamic model was derived by using the method of assumed modes as shown in Appendix C. Defining $X^T = [q \; \delta_1 \cdots \delta_m]$, where m is the number of flexible modes, and y as the vector of strain outputs, a linearized model is obtained in the following form

$$\begin{aligned} \dot{X} &= AX + bu \\ y &= CX \end{aligned} \qquad (2.48)$$

where u is the command torque input and

$$A = \begin{bmatrix} O_{n \times n} & I_{n \times n} \\ -M_0^{-1} K_0 & O_{n \times n} \end{bmatrix}, \quad b = M_0^{-1}[1 \; 0 \; \cdots \; 0]^T$$

$$C = \begin{bmatrix} 0_{m \times 1} & C_{\delta_{m \times m}} & O_{m \times n} \end{bmatrix}, \quad K_0 = \begin{bmatrix} 0_{1 \times 1} & 0_{1 \times m} \\ 0_{m \times 1} & K \end{bmatrix} \qquad (2.49)$$

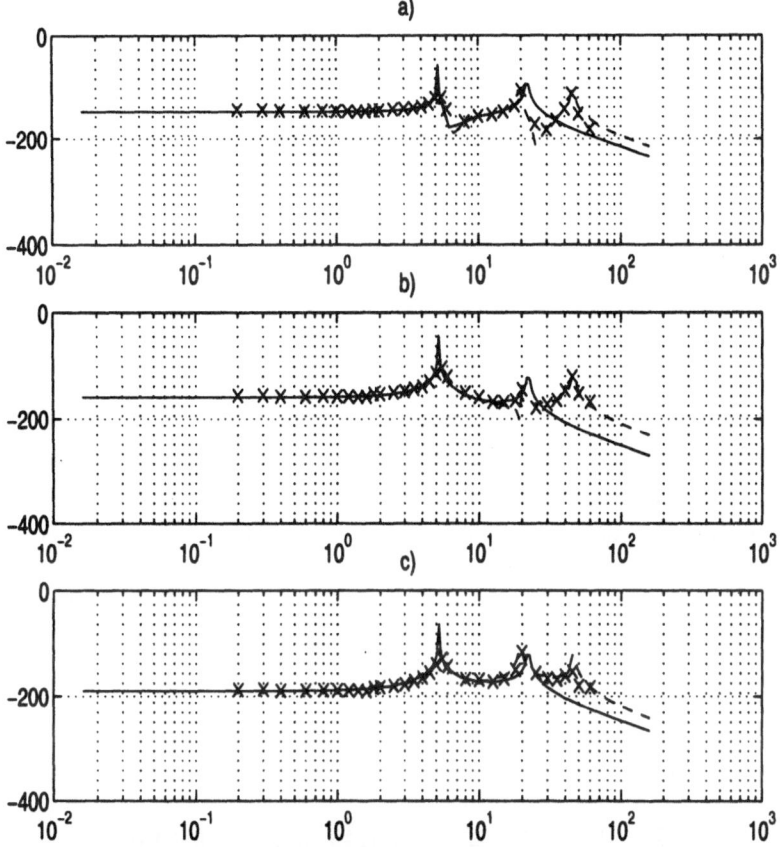

Figure 2.7. Comparison between the experimental (×) and analytical frequency responses for two (—) and three (− −) flexible modes when the strain gauge output is taken at   a) Point 1,   b) Point 2,   and   c) Point 3.   Horizontal axis: Frequency (Hz), Vertical axis: $(strain/u)$ dB .

where $n = m + 1$, $M_0 = M(q, \delta = 0)$, and $C_{\delta_{m \times m}}$ is obtained from (2.47). A sketch of the magnitudes of frequency responses obtained from this model for $m = 2$ and $m = 3$ along with the experimental results are given in Figure 2.7. It is observed that the experimental results do closely match the predicted results obtained from the model based on the assumed modes method. As the number of modes is increased from two to three, the higher frequency portions of the curve gets closer to the experimental data. Moreover, it was found that the clamped–mass shape functions resulted in a closer match than the clamped–free shape functions.

2.4.2. Implementation of the Control Law

The terms required by the controller are based on the model given in the Appendix C and are as follows

$$d(x_1, x_2, \dot{x}_2 \mid_{\varepsilon=0}, u_0, \dot{u}_0) = 3 \times 10^{-6} \dot{q}^2 v_0 - 37.23 \ddot{v}_0$$

$$\ddot{h}_{10} = \begin{bmatrix} -26.6512 v_0 \\ 0.0296 v_0 \end{bmatrix}, \quad h_{11} = \begin{bmatrix} -26.6482 v_1 \\ -0.2964 v_1 \end{bmatrix}$$

$$h_{12} = \begin{bmatrix} 19.9221 \ddot{v}_0 - 1011.8217 \dot{q}^2 v_0 - 199.7986 u_2 \\ -1.0965 \ddot{v}_0 + 0.0177 \dot{q}^2 v_0 - 0.2222 u_2 \end{bmatrix}.$$

For the open–loop flexible–link system the roots of the linearized zero dynamics are located at ± 15.4, $\pm j72.2$. The design matrices F, G, and M are selected so that the poles of A_η, given by (2.44), are in the left–half of the complex plane. Thus, choosing $P_{K_d} = \{-0.1 \pm j4.2, -0.1 \pm j1.0\}$ and $P_{L_d} = \{-0.2 \pm j8.4, -0.2 \pm 2.0\}$, results in

$$F = \begin{bmatrix} -0.3773 & 0.3773 & 1.0000 & 0 \\ 0.4227 & -0.4227 & 0 & 1.0000 \\ -101.6687 & 20.3974 & -1.6594 & 7.3035 \\ -16.0989 & -0.7886 & -0.2861 & 1.2594 \end{bmatrix}$$

$$G = 10^4 \times [0.0133 \quad -0.0149 \quad 3.2128 \quad 0.5014]$$

and

$$M = [0.0004 \quad -0.0006 \quad 0.0019 \quad -0.0083].$$

The other data parameters for this system are given as $\varepsilon = 0.03$, $K_p = 1$, $K_d = 2$, $A_1 = 2.5$, and $A_2 = 37$.

The differential equations corresponding to the dynamic control laws (2.35) and (2.41) have to be numerically solved for a digital implementation. The numerical solution has to be fast enough so that the results are computed and made available well before the end of each sampling period. The procedure adopted here is the *modified midpoint* method [74]. On average, this method requires 1.5 derivative evaluations per step as compared to the Runge–Kutta's 4 evaluations. Three steps were used during each sampling period. The implemented algorithm took 2.2 msec on the MC68030 Ironics processor board. Thus, a sampling frequency of 350 Hz was used. This rate was sufficient to allow computation of the control law as well as the data acquisition and trajectory generation tasks, while maintaining closed–loop system stability.

Due to mechanical imprecisions in the physical construction of the manipulator, the arm is not completely level in the horizontal plane. This can lead to considerable errors due to the fact that the magnitude of the required torque for control is small. The problem is resolved by noting that the gravity field produces a torque about the joint axis that may be expressed by $\tau_g = a + b\cos(q) + c\sin(q)$, where q is the joint angle and a, b and c are terms due to small offset angles. These terms are estimated by measuring the balancing torques at several joint angles and by using the least–squares algorithm. Thus, τ_g is added to the control torque to counterbalance the gravitational effects. The experimental results are shown in Figure 2.8, for the case when the proposed control strategy is applied, and in Figure 2.9, for the case when the fast control remains the same but a rigid body slow control is employed. The experimental results show improved tracking performance for the proposed method. The ratio of maximum tip position error to maximum arm deflection during the whole trajectory is found to be 0.67 for the proposed scheme and 1.39 for the slow rigid–based control method. The steady–state errors in both figures are a result of the small arm deflection that exists due to gravity effects. The maximum tip angular error of the simulated 2–link nonlinear system is about $0.005rad$ while it is $0.05rad$ for the experimental arm. It should be noted that a faster trajectory is used in the experimental case. On the other hand, it is worth emphasizing that the theoretical tracking error bound in Theorem 2.1 is derived subject to the absence of modeling imperfections, friction terms, actuator dynamics, discretization effects, sensor noise, higher frequency unmodeled dynamics, and computational delay of the control law. These factors are among the possible sources that contribute to the difference between the theoretical tracking error estimates and the actual tracking errors.

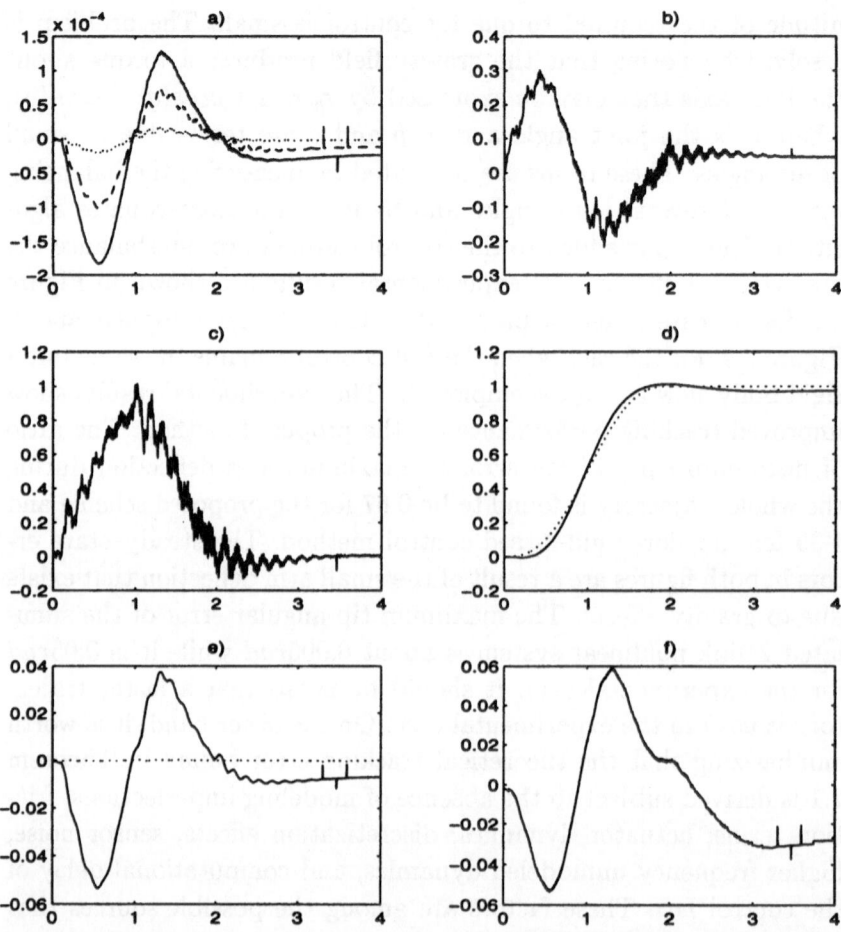

Figure 2.8. Experimental results for the proposed method (Horizontal axis: time (s)):   a) Strains at points 1 (—), 2 (--), and 3 (···) (m/m) b) Torque input (Nm)   c) Joint velocity (rad/s)   d) Tip trajectory (—) and desired tip trajectory (···) (rad)   e) Deflection modes δ_1 (—) and δ_2 (···) (m)   f) Tip position trajectory error, $y_r - y$ (rad).

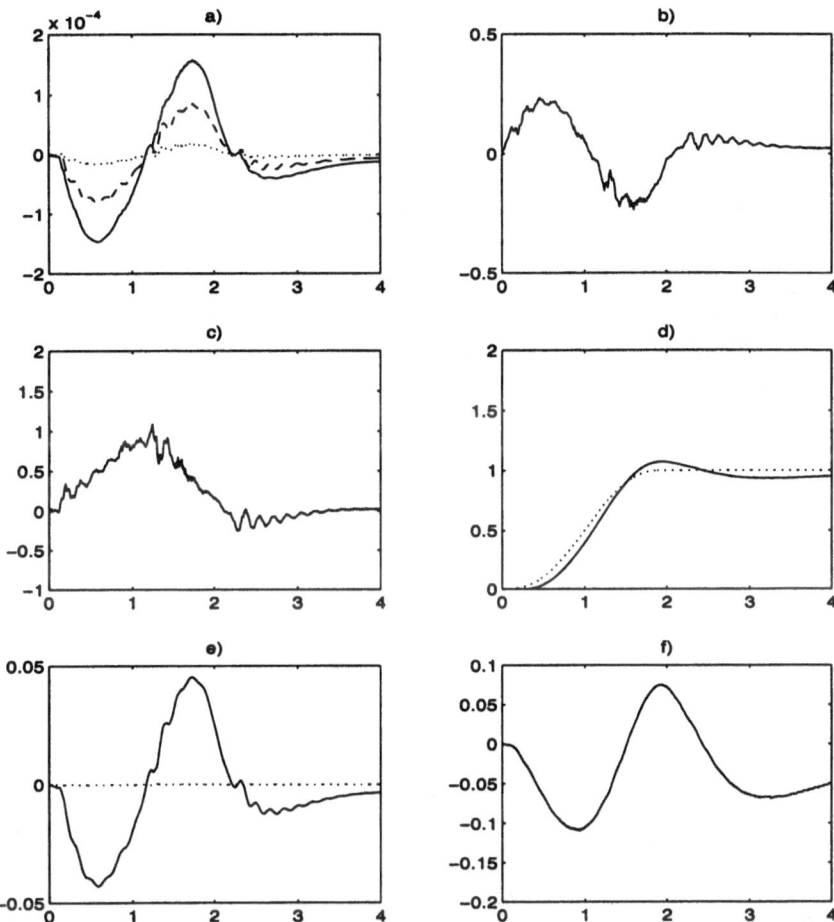

Figure 2.9. Experimental results for the slow control designed based on the rigid model plus the fast control (Horizontal axis: time (s)):   a) Strains at points 1 (—), 2 (—–), and 3 ($\cdots$) (m/m)   b) Torque input (Nm)   c) Joint velocity (rad/s) d) Tip trajectory (—) and desired tip trajectory ($\cdots$) (rad)   e) Deflection modes δ_1 (—) and δ_2 ($\cdots$) (m)   f) Tip position trajectory error, $y_r - y$ (rad).

2.5.   Conclusion

In this chapter, a control scheme was proposed for achieving greater accuracy for tip position tracking in structurally flexible robotic manipulators. Theoretical estimates show that the tracking errors converge to a residual set of $O(\varepsilon^2)$, and experimental results show that smaller tip position tracking errors are achieved compared to conventional algorithms in the literature. The only measurements required by the control law are joint positions, velocities and tip positions. This is an attractive feature from a practical implementation point of view as tip deflection rates are not directly measured, but tip positions can be readily measured by camera vision systems or strain gauge sensors. The control law is more complicated than its rigid counterpart; however the use of symbolic manipulation software and fast real–time control technology make the implementation of such a control law feasible.

3.

Decoupling Control

In this chapter we present an inverse dynamics control strategy to achieve sufficiently small tracking errors for a class of multi–link structurally flexible manipulators. This is done by defining new outputs near the end points of the arms as well as by augmenting the control inputs by terms which ensure stable operation of the closed–loop system under specific conditions. The controller is designed in a two–step process. First, a new output is defined such that the zero dynamics of the corresponding system are stable. Next, to ensure stable asymptotic tracking the control input is modified such that stable asymptotic tracking of the new output or approximate tracking of the actual output may be achieved. This is illustrated for the case of single– and two–link flexible manipulators.

3.1. Introduction

As discussed in Chapter 1, it is well known that the transfer function from the torque input to the tip position output of a single–link manipulator is, in general, non–minimum phase [36]. For a causal controller, the non–minimum phase property hinders perfect asymptotic tracking of a desired tip trajectory with a bounded control input. Thus, for a causal controller and perfect tracking, the flexible system should be minimum phase. The minimum phase property may be achieved by output re–definition, as done in [36], [37], [38], or by a redefinition of the output into *slow* and *fast* outputs as done in the previous chapter. In [36], the reflected tip position was proposed as the output for a single–link manipulator. In [37], the region of sensor and actuator locations for achieving the minimum–phase property for a single–link manipulator was investigated using a linear transfer function of the link. In [38], a region of outputs having the minimum phase property was given for a two–link manipulator. However, the approach is restricted since it consists of numerical calculations for a specific manipulator with two flexible modes.

In this chapter, after deriving the zero dynamics of a certain class of multi–link flexible manipulators, we will develop a control strategy based on input–output linearization of the flexible–link system. Modeling uncertainties have been taken into account, and it is further assumed that the vibrations are mainly lateral vibrations about the axis of rotation. In other words, for each link it is assumed that a considerable amount of potential energy is stored in the direction of bending corresponding to the axis of rotation of that link, and that the potential energies due to deflections in other directions are negligible. This may be achieved by proper mechanical structure design. A planar manipulator with rectangular cross sections in which the height to thickness ratio of each cross section is large is an example of such a system.

The results are applied to a two–link manipulator. In particular, regions corresponding to the minimum–phase property are obtained and compared under different load and damping conditions. The control strategy is also tested on flexible single– and two–link manipulators.

3.2. Input–output Linearization

The input–state map of flexible–link manipulators is not in general feedback linearizable [56]. However, the system is locally input–output linearizable. Input–output linearization in nonlinear systems theory is essentially based on the developments described in [18] and [19]. In order to apply this technique to flexible–link manipulators, let us first consider the dynamics of a multi–link flexible manipulator

[29] with structural damping added, i.e.,

$$M \begin{bmatrix} \ddot{q} \\ \ddot{\delta} \end{bmatrix} + \begin{bmatrix} f_1(q,\dot{q}) + g_1(q,\dot{q},\delta,\dot{\delta}) + E_1\dot{q} \\ f_2(q,\dot{q}) + g_2(q,\dot{q},\delta,\dot{\delta}) + E_2\dot{\delta} + K\delta \end{bmatrix} = \begin{bmatrix} u \\ 0 \end{bmatrix} \quad (3.1)$$

where q is the $n \times 1$ vector of joint variables, δ is the $m \times 1$ vector of deflection variables, f_1, f_2, g_1, and g_2 are the terms due to gravity, Coriolis, and centripetal forces, M is the positive–definite mass matrix, E_1 and E_2 are positive–definite damping matrices, K is the positive–definite stiffness matrix, and u is the $n \times 1$ vector of input torques (clamped mode shapes have been assumed). Let us define $H(q,\delta) = M^{-1}(q,\delta) = \begin{bmatrix} H_{11} & H_{12} \\ H_{21} & H_{22} \end{bmatrix}$. Then (3.1) can be written in the state–space form

$$\dot{x} = f(x) + g(x)u \quad (3.2)$$

where $x^T = \begin{bmatrix} q^T & \delta^T & \dot{q}^T & \dot{\delta}^T \end{bmatrix}$,

$$f(x) = \begin{bmatrix} \dot{q} \\ \dot{\delta} \\ -H_{11}(f_1 + g_1 + E_1\dot{q}) - H_{12}(f_2 + g_2 + K\delta + E_2\dot{\delta}) \\ -H_{21}(f_1 + g_1 + E_1\dot{q}) - H_{22}(f_2 + g_2 + K\delta + E_2\dot{\delta}) \end{bmatrix},$$

$$g(x) = \begin{bmatrix} O_{(m+n)\times n} \\ H_{11}(q,\delta) \\ H_{21}(q,\delta) \end{bmatrix}.$$

Following [38], since the beam deflection is usually small with respect to link length, we have from Figure 3.1,

$$y_i = q_i + \alpha_i d_{ie}/l_i, \quad i = 1,2,\cdots,n. \quad (3.3)$$

In (3.3), α_i is a variable which takes values between -1 and $+1$, with $\alpha_i = 1, 0, -1$ corresponding to tip, joint angle, and reflected tip positions respectively.

The tip deflection d_{ie} can be written as

$$d_{ie} = \sum_{j=1}^{n_i} \Phi_{ij}(l_i)\delta_{ij} \quad (3.4)$$

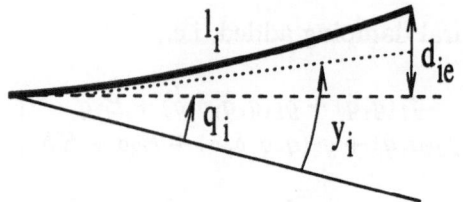

Figure 3.1. The output of link i (y_i).

where Φ_{ij} is the j-th mode shape function of the i-th link and δ_{ij} is the j-th mode of the i-th link. Thus, for the output vector, we have (see also Appendix B)

$$y = q + \Psi_{n \times m} \delta \qquad (3.5)$$

where

$$
\begin{aligned}
\Psi_{n \times m} &= \begin{bmatrix} v_1^T & 0^T & \cdots & 0^T \\ 0^T & v_2^T & \cdots & 0^T \\ \vdots & & \ddots & \vdots \\ 0^T & 0^T & \cdots & v_n^T \end{bmatrix} \\
(v_i^T &= \frac{\alpha_i}{l_i} \begin{bmatrix} \Phi_{i1}(l_i) & \cdots & \Phi_{in_i}(l_i) \end{bmatrix}, i = 1, \cdots, n) \\
y^T &= \begin{bmatrix} y_1 & \cdots & y_n \end{bmatrix} \\
\delta^T &= \begin{bmatrix} \Delta_{n_1}^T & \Delta_{n_2}^T & \cdots & \Delta_{n_n}^T \end{bmatrix} \qquad (3.6)
\end{aligned}
$$

with Δ_{n_i} being the vector of the deflection variables of link i, defined as

$$\Delta_{n_i}^T = \begin{bmatrix} \delta_{i1} & \cdots & \delta_{in_i} \end{bmatrix}. \qquad (3.7)$$

Now consider system (3.2), with the output defined by (3.5). To perform input–output linearization on this system we take time derivatives until the inputs appear. In our case, two differentiations are required, after which all inputs appear simultaneously. After some manipulations we have,

$$\ddot{y} = a(\alpha, x) + B(\alpha, q, \delta)u \qquad (3.8)$$

where

$$\alpha^T = \begin{bmatrix} \alpha_1 & \cdots & \alpha_n \end{bmatrix},$$
$$B(\alpha, q, \delta) = H_{11} + \Psi_{n \times m} H_{21},$$

$$a(\alpha, x) = -(H_{11} + \Psi H_{21})(f_1 + g_1 + E_1\dot{q}) - (H_{12} + \Psi H_{22})$$
$$\times (f_2 + g_2 + K\delta + E_2\dot{\delta}). \quad (3.9)$$

Now suppose that α and q_r have been selected such that $B(\alpha, q_r, 0)$ is nonsingular (q_r denotes the desired reference trajectory to be tracked by y). Then continuity implies that, on a finite domain around $x_r^T = [q_r^T \ 0 \ \dot{q}_r^T \ 0]$, taking u as

$$u = B^{-1}(\alpha, q, \delta)(v - a(\alpha, x)) \quad (3.10)$$

results in

$$\ddot{y} = v \quad (3.11)$$

which is an input–output linearized system with a new input vector v. Note that the dimension of the unobservable dynamics is $2m$. Now consider the state transformation

$$z = T(x) = D_T x \quad (3.12)$$

where $z := [z_o^T \ z_u^T]$, and

$$z_o^T = [z_{o1}^T \ z_{o2}^T]$$

$$z_u^T = [z_{u1}^T \ z_{u2}^T]$$

$$D_T = \begin{bmatrix} I_{n\times n} & \Psi_{n\times m} & O_{n\times n} & O_{n\times m} \\ O_{n\times n} & O_{n\times m} & I_{n\times n} & \Psi_{n\times m} \\ O_{m\times n} & I_{m\times m} & O_{m\times n} & O_{m\times m} \\ O_{m\times n} & O_{m\times m} & O_{m\times n} & I_{m\times m} \end{bmatrix}.$$

Because of the nonsingularity of D_T, (3.12) is a global diffeomorphism, and will transform (3.2) with output given by (3.5) into

$$\begin{aligned} \dot{z}_{o1} &= z_{o2} \\ \dot{z}_{o2} &= a(\alpha, x) + B(\alpha, q, \delta)u \\ \dot{z}_u &= C(x) + D(x)u \\ y &= z_{o1} \end{aligned} \quad (3.13)$$

where $C(x)$ and $D(x)$ are matrices corresponding to $f(x)$ and $g(x)$. To find the zero dynamics, z_{o1} and z_{o2} are set identically to zero, which after some manipulations leads to an explicit relationship for the zero dynamics, i.e.

$$
\begin{aligned}
\dot{z}_{u1} &= z_{u2} \\
\dot{z}_{u2} &= [-H_{22} + H_{21}(H_{11} + \Psi H_{21})^{-1}(H_{12} + \Psi H_{22})]\,|_{(w_1,w_3)} \\
&\quad \times\ [f_2(w_1,w_2) + g_2(w_1,w_2,w_3,w_4) + Kz_{u1} + E_2 z_{u2}] \quad (3.14)
\end{aligned}
$$

where

$$
w_1 = -\Psi z_{u1},\ w_2 = -\Psi z_{u2},\ w_3 = z_{u1},\ w_4 = z_{u2}. \tag{3.15}
$$

Linearizing (3.14) about the equilibrium point $z_{u1}, z_{u2} = 0$ gives

$$
\begin{aligned}
\dot{z}_{u1} &= z_{u2} \\
\dot{z}_{u2} &= [-H_{22} + H_{21}(H_{11} + \Psi H_{21})^{-1}(H_{12} + \Psi H_{22})]\,|_{z_{u1},z_{u2}=0} \\
&\quad \times\ (Kz_{u1} + E_2 z_{u2}). \tag{3.16}
\end{aligned}
$$

The following result is now concluded from the above discussion:

Condition for Minimum–phase Behavior:

Let the vector α, and the matrices $H(0,0), K$, and E_2 be such that the matrix

$$
A(\alpha) = \begin{bmatrix} O & I \\ -P_0 K & -P_0 E_2 \end{bmatrix} \tag{3.17}
$$

with P_0 given by

$$
P_0 = [H_{22} - H_{21}(H_{11} + \Psi H_{21})^{-1}(H_{12} + \Psi H_{22})]_{(0,0)} \tag{3.18}
$$

is a Hurwitz matrix. Then the origin of (3.16), and hence (3.14), is locally asymptotically stable, and the original nonlinear system is locally minimum phase. This result follows by noting that the eigenvalues of A are the modes of the linearized zero dynamics of the system. It is interesting to note that $\alpha = 0$ (joint position output) guarantees P_0 to be a positive definite matrix, which makes A a Hurwitz matrix. This can be shown by using the Lyapunov function candidate $V = z_{u1}^T K z_{u1} + z_{u2}^T P_0^{-1} z_{u2}$ for (3.16) and applying LaSalle's theorem.

Now suppose that, due to modeling errors and truncation of modes, we cannot exactly get u given by (3.10). Furthermore, since the damping terms E_1 and E_2 are not commonly modeled exactly, we include them in the uncertainties. Thus in the previous relations, it suffices to put $E_1 = E_2 = 0$ and account for them in the uncertainties. Let us define

$$u = \hat{B}^{-1}(\alpha, q, \delta)(v - \hat{a}(\alpha, x)) + K_\delta(q)\delta + K_{\dot{\delta}}(q)\dot{\delta} \qquad (3.19)$$

where $B = \hat{B} - \Delta B$ and $a = \hat{a} - \Delta a$, with ΔB and Δa representing the uncertainties in B and a respectively [1]. Then, from (3.8) we have

$$\ddot{y} = B\hat{B}^{-1}v + (I - B\hat{B}^{-1})a - B\hat{B}^{-1}\Delta a + BK_\delta(q)\delta + BK_{\dot{\delta}}(q)\dot{\delta}. \qquad (3.20)$$

In view of (3.13) with u given by (3.19) and choosing

$$v = \ddot{y} + K_p e + K_d \dot{e} \qquad (3.21)$$

where $e = y_r - y$ (y_r is the desired trajectory for the output), yields

$$\dot{E} = A_E E + d_E(\alpha, x, , t) \qquad (3.22)$$

with

$$
\begin{aligned}
d_E(\alpha, x, t) &= (I - B\hat{B}^{-1})(K_p e + K_d \dot{e} + \ddot{y}_r - a) \\
&+ B\hat{B}\Delta a - BK_\delta(q)\delta + K_{\dot{\delta}}(q)\dot{\delta} \\
E^T &= \begin{bmatrix} e^T & \dot{e}^T \end{bmatrix} \\
A_E &= \begin{bmatrix} 0 & I \\ -K_p & -K_d \end{bmatrix}.
\end{aligned}
\qquad (3.23)
$$

Similarly (3.13) can be written in terms of the new u, that is

$$\dot{\Delta} = A_\Delta(q)\Delta + \begin{bmatrix} 0 \\ G_\Delta(x,t) \end{bmatrix} \qquad (3.24)$$

where

$$
\begin{aligned}
G_\Delta(x,t) &= H_{210}\hat{B}_0^{-1}(\ddot{y}_r + K_p e + K_d \dot{e}) + H_{210}(-\hat{B}_0^{-1}\hat{a}_0 + B_0^{-1}a_0) \\
&+ O(\delta^2, q, \dot{q}) \\
A_\Delta(q) &= \begin{bmatrix} 0 & I \\ -P(q)K - H_{210}K_\delta(q) & -H_{210}K_{\dot{\delta}}(q) \end{bmatrix} \\
\Delta^T &= [\delta^T \ \dot{\delta}^T]
\end{aligned}
\qquad (3.25)
$$

---

[1] Note that $\hat{B}$ is at our disposal. For example, it may be obtained from a model with two flexible modes while the actual plant may be described by three flexible modes.

and $O(\delta^2, q, \dot{q})$ indicates that the remaining terms are of order δ^2, and $H_{210} = H_{21}(q, 0)$ (the same notation is adopted for the other H_{ij} $(i, j = 1, 2)$ submatrices in the sequel). Furthermore, matrices $K_\delta(q)$ and $K_{\dot{\delta}}(q)$ are selected such that $A_\Delta(q)$ is a Hurwitz matrix for the range in which q is varied and

$$P(q) = [H_{220} - H_{210}(H_{110} + \Psi H_{210})^{-1}(H_{120} + \Psi H_{220})]. \quad (3.26)$$

Noting that A_E and A_Δ are Hurwitz matrices, it may then be shown (using a Lyapunov analysis) that the trajectories of the closed–loop system given by (3.22) and (3.24) converge to a residual set with small tracking errors e and $\dot{e}$ and bounded δ and $\dot{\delta}$ provided that certain conditions are satisfied. The following theorem summarizing the above results may then be stated.

Theorem 3.1 *Let the control law (3.19) be applied to the original nonlinear system (3.1) with $\hat{B}$ nonsingular on the domain of interest, then assuming that the desired trajectories and their time derivatives (at least up to order 2) are continuous and bounded, it follows that the trajectories of E, $\epsilon_1 \Delta$ (ϵ_1 is a positive number typically less than one), starting from a specific set, converge to a small residual set provided that certain norm conditions in a bounded region of the state space of E and Δ, containing the origin are satisfied.*

Proof: The above result is proved in Appendix A.2.

Remark 1: Singular B Matrix
Here we have assumed that $B(\alpha, q, \delta)$ is nonsingular. In case B is singular the decoupling method described above cannot be achieved by static state feedback. However it may still be possible to find a dynamic compensator of the form

$$\begin{aligned} \dot{z} &= d_c(x, z) + E_c(x, z)U \\ u &= f_c(x, z) + G_c(x, z)U \end{aligned} \quad (3.27)$$

with $z \in R^{n_c}$, $U \in R^n$, such that the extended system described by (3.8) and (3.27) with output given by (3.5) is decoupled from U to y. The theoretical developments for general affine nonlinear systems are given in [68]. From a practical point of view in our application, employing this method may have the drawback of requiring acceleration (or higher derivative) measurements.

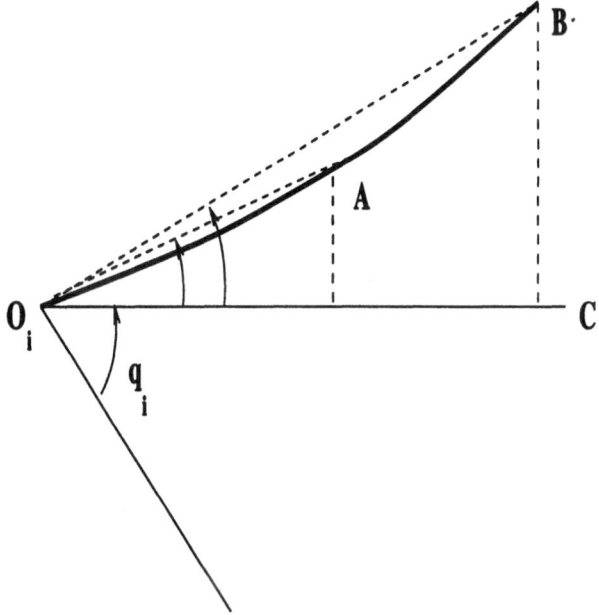

Figure 3.2. Tip position and redefined outputs of link i.

Remark 2: Tip-position Tracking Errors
In Theorem 3.1, we mention redefined output tracking errors which have been denoted by e_1 and e_2. The redefined outputs are near the tip positions. The controller tries to achieve small tracking errors at these points. The additional tracking errors at the tip positions are therefore a result of the effect of moving the nearby redefined outputs along the desired trajectories. An approximate relationship for the additional tracking error between the redefined output of link i and its tip position can be found as follows. Refer to Figure 3.2 where point A corresponds to the redefined output and B corresponds to the tip–position of link i. Thus the additional error corresponds to the angle $\angle BO_iA$ in this figure which can be obtained from

$$\angle BO_iA = tan^{-1}\frac{\sum_k \phi_{ikt}\delta_k}{l_i} - tan^{-1}\frac{\sum_k \phi_{ik\alpha_i}\delta_k}{\alpha l_i} \tag{3.28}$$

where ϕ_{ikt} and $\phi_{ik\alpha_i}$ denote the kth shape function of link i evaluated at the tip position and the redefined output of link i, respectively. Assuming that the deflections are small compared to the length of the link, (3.28) can be written as

$$\angle BO_iA = \sum_k \frac{\alpha_i\phi_{ikt} - \phi_{ik\alpha_i}}{\alpha_i l_i}\delta_k. \tag{3.29}$$

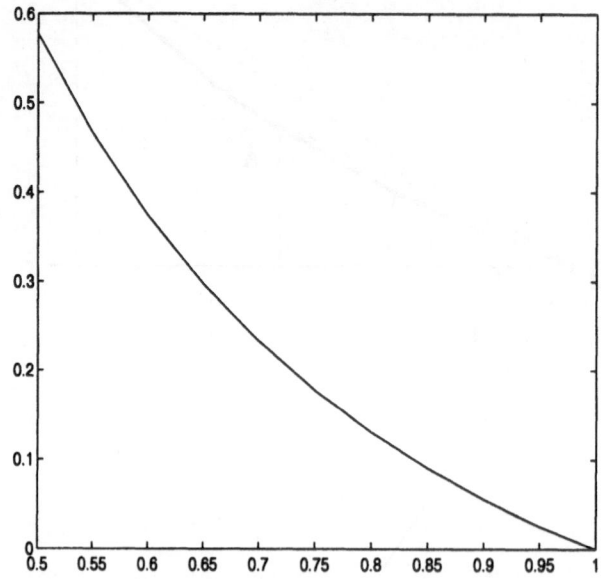

Figure 3.3. Variation of the approximate relative error $\frac{\angle BO_iA}{\angle AO_iC}$ (vertical axis) with the α_i of the redefined output (horizontal axis).

Now if (3.29) is normalized to the deflection of the link at the redefined output (measured from the rigid body angle q_i) we have

$$\frac{\angle BO_iA}{\angle AO_iC} = \frac{\sum_k(\alpha_i\phi_{ikt} - \phi_{ik\alpha_i})\delta_k}{\sum_k \phi_{k\alpha_i}\delta_k}. \tag{3.30}$$

Since the desired trajectories are usually smooth, the first flexural modes are excited more than the rest, i.e., $\delta_1 \gg \delta_2 \gg \cdots$. Therefore an approximation to (3.30) can be found as

$$\frac{\angle BO_iA}{\angle AO_iC} \approx \frac{\alpha_i\phi_{ilt} - \phi_{1\alpha_i}}{\phi_{1\alpha_i}}. \tag{3.31}$$

3.2.1. Derivation of H(0,0)

It is noted that in order to use (3.17)–(3.18), $H(0,0)$ (or $M(0,0)$) should be known. Fortunately, since only M at $q = 0$ and $\delta = 0$ is required, the calculation of M is greatly simplified and a general method can be established for deriving $M(0,0)$ with any number of modes. This was done for a planar two–link manipulator used in our case study. To do so, it is sufficient to find the kinetic energy $\mathcal{K}_e$ of the system when it passes through $q = 0$ and $\delta = 0$ and then use $\mathcal{K}_e(0,0) = \frac{1}{2} \begin{bmatrix} \dot{q}^T & \dot{\delta}^T \end{bmatrix} M(0,0) \begin{bmatrix} \dot{q} \\ \dot{\delta} \end{bmatrix}$ to find $M(0,0)$.

3.2.2.   A Model for the Damping Term E_2

In this section we establish a model for the term E_2 introduced in (3.1) and used in (3.17). Considering (3.1) and eliminating $\ddot{q}$ from the equations, we get

$$N\ddot{\delta} + E_2\dot{\delta} + K\delta - M_{21}M_{11}^{-1}(V_1 + E_1\dot{q}) + V_2 = -M_{21}M_{11}^{-1}u \quad (3.32)$$

where V_1 and V_2 are the terms due to Coriolis and centripetal forces, and

$$N(q,\delta) = M_{22}(q,\delta) - M_{21}(q,\delta)M_{11}^{-1}(q,\delta)M_{12}(q,\delta). \quad (3.33)$$

Considering the dynamics of flexible modes ((3.32)) at $q, \dot{q} = 0$, and linearizing this equation about $\delta, \dot{\delta} = 0$ gives

$$N_o\ddot{\delta} + E_2\dot{\delta} + K\delta = f := -M_{21}M_{11}^{-1}u \quad (3.34)$$

where f is the total resulting forcing function and $N_o = N(0,0)$. Physically, (3.34) describes the dynamics of the flexible modes of the system when all the joints are locked at $q = 0$. Note that the matrices in (3.17)–(3.18) were evaluated at $q = 0$ and $\delta = 0$. The damping term can now be postulated as $E_2 = \alpha_N N_o + \beta_K K$ where α_N and β_K are constant positive scalars. This model is known as *Rayleigh Damping* in vibration theory (see e.g. [33]). Thus the damping factor of each mode is given [33] by $\xi_i = \frac{\alpha_N + \beta_K \omega_i^2}{2\omega_i}$, $i = 1, 2, \cdots, m$, where ω_i^2 is the i-th eigenvalue of $N_o^{-1}K$.

3.3.   Case Studies

3.3.1.   Regions having the Minimum–phase Property for a Two–link Manipulator

A planar two–link manipulator is studied in this section and the regions of output locations for achieving the minimum–phase property are obtained. The manipulator consists of two uniform bars with

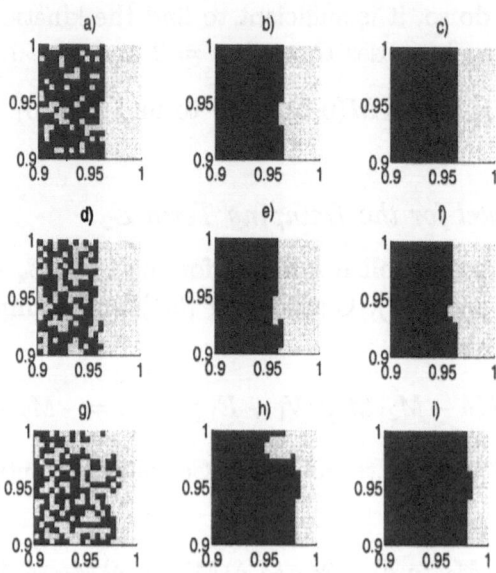

Figure 3.4. Regions of outputs for minimum–phase behavior (dark areas). Horizontal axis: α_1, Vertical axis: α_2. (a, b, c) Regions for $\xi_1 = 10^{-14}, 10^{-7}, 10^{-2}$ respectively when three flexible modes are used ($M_1 = M_2 = 0.52kg$, $J_1 = J_2 = 0.17kgm^2$).   (d, e, f) Same as a–c when four flexible modes are used.   (g, h, i) Same as d–f but with different loading, i.e., $M_1 = M_2 = 1.55kg$ and $J_1 = J_2 = 0.51kgm^2$.

rectangular cross sections and considerable flexibility with the following numerical data

$$l_1 = l_2 = 0.7m, \ A_1(x_1) = A_2(x_2) = 7.44cm \times 0.46cm,$$

$$\rho = 2700kg/m^3 \ (6061 Aluminum), \ M_1 = M_2 = 0.52kg,$$

$$E = 69.3 \times 10^9 N/m^2, J_1 = J_2 = 0.17kgm^2$$

where l_1 and l_2 are link lengths, A_1 and A_2 are cross sectional areas, E and ρ are modulus of elasticity and mass density, and M_1, M_2, J_1 and J_2 are masses and mass moments of inertia at the end points of each link.

The damping model was chosen as discussed in the previous section. It was assumed that the damping ratio of the first mode, ξ_1, is given, and α_N, β_K contribute equally to this damping ratio. The extreme cases where either α_N or β_K are zero were also considered. The results were similar. Here, results are presented only for equal contribution of α_N and β_K. The vector α in (3.9) has the form $\alpha^T = [\alpha_1 \alpha_2]$.

The terms α_1 and α_2 were varied from -1 to $+1$ and the matrix A in (3.17) was tested for eigenvalues in the left half plane. Figure (3.4) shows the results for the region near the tip positions from $\alpha = 0.9$ to $\alpha = 1$. Three flexible modes were used in Figures 3.4a–3.4c, but for the rest, four flexible modes were used. The damping ratio was chosen as 10^{-14}, 10^{-7}, and 10^{-2}. When the damping ratio is very small, i.e. 10^{-14}, the regions of minimum–phase behavior are not reliable. A change in load configuration (Figure 3.4e–3.4i), or the number of flexible modes will change this behavior. However, note that the damping ratio (ξ_1) is of the order of 0.01 for commonly used metals. It is also interesting to note that the behavior does not change much when we change ξ_1 from 10^{-7} to 10^{-2}. Moreover, there are points (e.g. $\alpha_1 = 0.9$, $\alpha_2 = 1$) which will preserve the minimum-phase property in spite of changes in load configuration (M_1, M_2, J_1, J_2), damping ratio (ξ_1), and the number of flexible modes. The discontinuities in Figure 3.4 are as the result of the resolution by which α is varied.

3.3.2. Inverse–dynamics Control

A Single Flexible Arm
The control law developed in the previous sections was applied to a single–link flexible Aluminum arm with the following data

$$l = 1.3m, \ A = 8cm \times 1.5mm, \ M_p = 1kg, \ J_p = 0.002kgm^2,$$

$$J_h = 3 \times 10^{-5}kgm^2,$$

where l, A, M_p, J_p, J_h denote the length, payload mass, payload inertia and hub inertia respectively. The α–vector is now a scalar, and it was selected as $\alpha = 0.85$ designed based on a plant model with three modes. The control law (3.19) was then designed based on a model with two flexible modes with $K_p = 9$ and $K_d = 6$. In this way the performance of the controller is tested for higher frequency unmodeled dynamics. Figure 3.5 illustrates the simulation results. It is interesting to note that to achieve tip position tracking the hub angle has to evolve in an oscillatory fashion. In this case, the closed–loop system was stable for $K_\delta = K_{\dot\delta} = 0$, but this is not the case in general, as discussed in the next example.

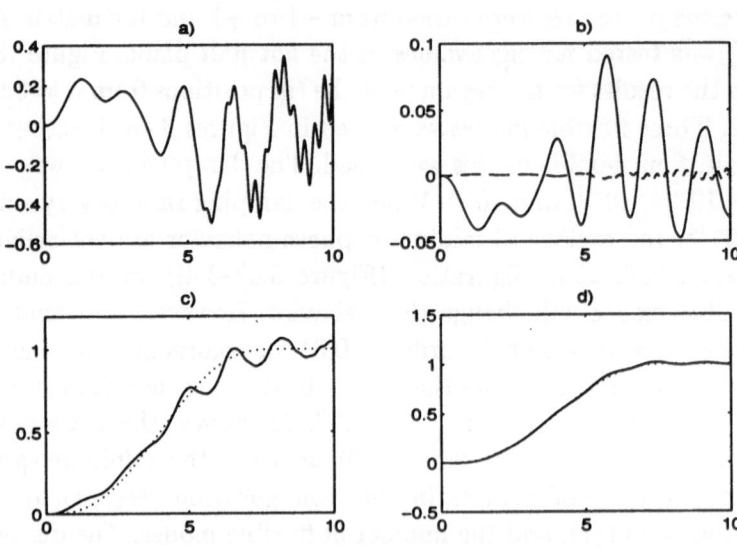

Figure 3.5. Simulation results for a single–link arm using inverse dynamics. (a) Input torque (Nm) vs. time (s)   (b) Flexible modes (m) (first mode:–, second mode:– –, third mode: not shown)   (c) Joint position (–) and reference trajectory (...) in radians   (d) Tip position (–) and reference trajectory (...) in radians.

A Two–link Planar Manipulator

A two–link planar manipulator is considered in which the first link is rigid and the second link is flexible. In this way significant non-linearities are introduced in the dynamic equations compared to the single–link case. The two–link data are

$$l_1 = 0.2m, \; l_2 = 1.3m, \; A_1 = 4.8387 \times 10^{-3} m^2, \; A_2 = 0.9975 \times 10^{-4} m^2$$

$$\rho = 7860 kg/m^3 \; (Steel), \; M_1 = M_2 = 1kg,$$

$$E = 206 \times 10^9 N/m^2, \; J_1 = J_2 = 0.002 kgm^2, \; J_h = 3 \times 10^{-5} kgm^2.$$

The two–link system is modeled by the assumed modes method with two flexible modes. Further E_1 and E_2 are assumed to be zero. Using (3.17), α was chosen to be close to 1 and such that the zeros are purely imaginary. This gave a critical value of $\alpha = 0.943$. Then $K_\delta(q)$ and $K_{\dot{\delta}}(q)$ were chosen such that at each point q (the second joint position variable), the matrix $A_\Delta(q)$ in (3.25) is a Hurwitz matrix. This was achieved by solving a pole placement problem at ten points and using linear interpolation to obtain $K_\delta(q)$ and $K_{\dot{\delta}}(q)$ at other points. It is important to choose the eigenvalues of $A_\Delta(q)$ close to the $j\omega$ axis so that small $K_\delta(q)$ and $K_{\dot{\delta}}(q)$ are obtained for better

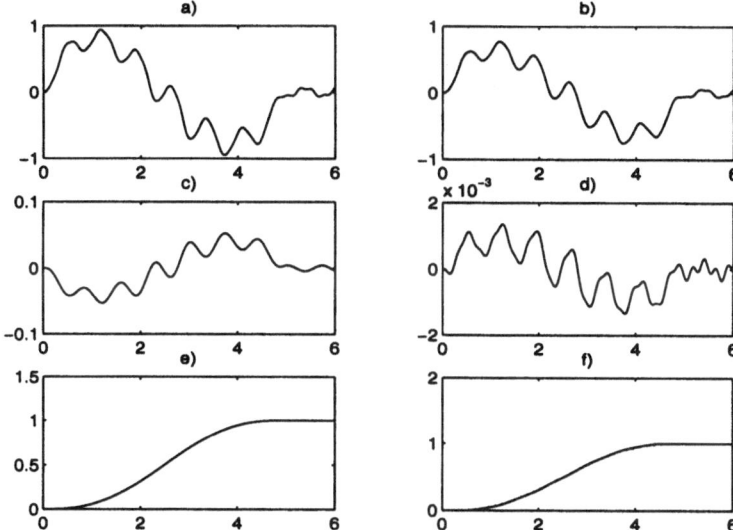

Figure 3.6. Simulation results for a two-link manipulator ($x - axis$: time (s)). (a) First actuator torque (Nm)   (b) Second actuator torque (Nm)   (c) First flexible mode (m)   (d) Second flexible mode (m)   (e) First tip position trajectory (–) and reference trajectory (...) in radians   (f) Second tip position trajectory (–) and reference trajectory (...) in radians.

robustness (see Appendix A.2). Setting these gains to zero resulted in an unstable system. The maximum torque required for tracking was about 5 Nm which was reduced to less than 1 Nm when α was reduced to 0.9. The results for $\alpha = 0.9$ are shown in Figure 3.6.

3.4. Conclusion

The control strategy discussed in this chapter yields sufficiently small tip–position errors and good robustness properties. This may be attributed to the decoupling effect of the input–output linearization technique. The control strategy was designed for a special class of manipulators in which the major deflection coordinates are in the same direction as the joint coordinates. A possible solution to the more general case is either through proper mechanical structure design or by introducing extra control inputs to affect flexibilities in other directions. From the two–link example it was also found that controlling the flexibility effects of the second link by the input control of the first link is hardly ever achievable and stability problems may arise. Thus the solution to the more general case may be offered by adding extra control inputs and better mechanical design. A drawback of this strategy may, however, be the requirement for full measurement of the states. Practically, joint positions and their rates as well as deflection variables (δ) are measurable but $\dot{\delta}$ is not directly measurable and should be estimated or reconstructed from deflection variable measurements (see e.g. [55]).

4.

Observer–based Decoupling Control

In this chapter, we focus on the design of an observer–based inverse dynamics control strategy that results in sufficiently small tip–position tracking errors while maintaining robust closed–loop performance for a class of multi–link structurally flexible manipulators. The control design is essentially based on the method described in Chapter 3. As part of the control design, a nonlinear observer is introduced to estimate the rates of change of flexible modes. By a proper choice of control and observer gains, the error dynamics are guaranteed to consist of a stable linear part plus a bounded perturbation term that results in asymptotically stable observation and closed–loop system stability with small tip position tracking errors. Experimental results are given for the case of a two–link flexible manipulator that further confirm the theoretical and simulation results.

4.1. Introduction

Accurate knowledge of state variables is required by many advanced control algorithms for flexible multi–link robots, e.g. see [57], [8], [79]. It is possible to measure joint positions, velocities, and flexible modes of manipulators using shaft encoders, tachometers, and strain gauges [64], respectively. However measuring rates of change of flexible modes cannot be easily or accurately accomplished. One method is to integrate the outputs of accelerometers installed along the arms or to use analog differentiation of the deflection variables [55]. The former approach may be restricted from economic considerations and the latter because of noise problems. Thus a nonlinear state observer is desirable in these circumstances. Several authors have studied the development of nonlinear observers in the general context of nonlinear systems or specifically intended for (mainly) rigid robot manipulators [60], [61], [62], [63]. In [58] the so called pseudo–linearization technique was used where the nonlinear robot dynamics are transformed into a linear model by a nonlinear state–space change of coordinates. Sliding techniques were introduced in [60], [61] where the attractive manifold concept was employed.

The organization of this chapter is as follows. First, the control strategy described in Chapter 3 is employed to choose points near the tip outputs such that stable zero–dynamics are achieved. As before, it is assumed that the vibrations are mainly lateral vibrations about the axis of each link. Second, an observation strategy is developed by studying certain characteristics of the dynamics of flexible modes of structurally flexible manipulators. In particular, the observer requires that joint angles and velocities as well as flexible modes are available in order to estimate the rates of change of flexible modes. It is also shown that sliding observer techniques such as those in [60], [61] can be easily incorporated in the design. The observation strategy is quite general and is applicable even if the above assumption regarding the arm shapes is relaxed.

A closed–loop stability analysis is performed and conditions for achieving stable closed–loop behavior are stated. The theoretical developments are further enhanced by experimental studies for a two–link flexible manipulator with promising results. In particular, stable closed–loop performance with small tip position tracking errors is achieved with readily available and economic sensor equipment. Furthermore, relatively large control gains can be used, resulting in reduced closed–loop system sensitivity that would otherwise not be achieved by conventional methods.

4.2.   Inverse–dynamics Control

Consider the control strategy discussed in Chapter 3. For convenience let us again re–write the dynamics of a flexible multi–link system given by (3.1) without the damping coefficients, i.e.,

$$
M(q, \delta)
\begin{bmatrix} \ddot{q} \\ \ddot{\delta} \end{bmatrix}
+
\begin{bmatrix} f_1(q, \dot{q}) + g_1(q, \dot{q}, \delta, \dot{\delta}) \\ f_2(q, \dot{q}) + g_2(q, \dot{q}, \delta, \dot{\delta}) + K\delta \end{bmatrix}
=
\begin{bmatrix} u \\ 0 \end{bmatrix}.
$$

$$(4.1)$$

Now, defining the output vector (see Chapter 3), i.e.,

$$
y = q + \Psi_{n\times m}(\alpha)\delta
\tag{4.2}
$$

where $\Psi(\alpha)$ is a matrix depending on modal shape functions and the vector $\alpha^T = [\alpha_1 \cdots \alpha_n]$ defines physical output locations on the links for achieving stable zero–dynamics [69]. The input–output description of (4.1) with the output described by (4.2) is then obtained by differentiating the output vector y with respect to time until the input vector appears, which is given by

$$
\ddot{y} = a(\alpha, x) + B(\alpha, q, \delta)u
\tag{4.3}
$$

where $B(\alpha, q, \delta) = H_{11} + \Psi_{n\times m}H_{21}$ and

$$
\begin{aligned}
a(\alpha, x) &= -(H_{11} + \Psi H_{21})(f_1 + g_1) \\
&\quad - (H_{12} + \Psi H_{22})(f_2 + g_2 + K\delta).
\end{aligned}
\tag{4.4}
$$

Now let us define a finite domain around the desired reference trajectory $q_r, \dot{q}_r$ given by

$$
\Omega_r = \{x: \ |q - q_r| < \kappa_1, \ |\dot{q} - \dot{q}_r| < \kappa_2, \ |\delta| < \kappa_3, \ |\dot{\delta}| < \kappa_4\}
\tag{4.5}
$$

where κ_i $(i = 1, \cdots, 4)$ are some positive bounds. Also assume that $B(\alpha, q, \delta)$ is nonsingular in Ω_r. This is a controllability like assumption and is guaranteed to hold when for instance $\alpha = 0$. Now, let u take the following form

$$
u = B^{-1}(\alpha, q, \delta)(v - a(\alpha, x)) + K_\delta(q)\delta + K_{\dot{\delta}}(q)\dot{\delta}
\tag{4.6}
$$

where $K_\delta(q)$ and $K_{\dot{\delta}}(q)$ are gain matrices that are to be specified later and intended to make $A_\Delta(q)$ given by (4.24) a Hurwitz matrix. It then follows from (4.3) that

$$
\ddot{y} = v + BK_\delta(q)\delta + BK_{\dot{\delta}}(q)\dot{\delta}
\tag{4.7}
$$

which is an input–output linearization of the system when $K_\delta(q)$ and $K_{\dot\delta}(q)$ are zero. As discussed later, these terms are added to enhance robustness and should usually be selected sufficiently small. In the above formulations, it is assumed that $\dot\delta$ is available. This assumption is relaxed later on by replacing $\dot\delta$ with its estimation from a nonlinear observer.

4.3.   Observer Design

In this section, our proposed observation strategy is introduced. Three cases are considered as described below.

4.3.1.   Full–order Observer

Consider the dynamics of a multi–link flexible manipulator given by (4.1). Defining $\delta_1 = \delta$ and $\delta_2 = \dot\delta$, the dynamics of flexible modes are written as

$$
\begin{aligned}
\dot\delta_1 &= \delta_2 \\
\dot\delta_2 &= -H_{21}(q,\delta_1)(f_1(q,\dot q) + g_1(x)) \\
&\quad - H_{22}(q,\delta_1)(f_2(q,\dot q) + g_2(x)) + K\delta_1) + H_{21}(q,\delta_1)u \quad (4.8)
\end{aligned}
$$

where $x^T = [q^T \ \dot q^T \ \delta_1^T \ \delta_2^T]$ as before. Let us choose the following structure for the observer dynamics

$$
\begin{aligned}
\dot{\hat\delta}_1 &= \hat\delta_2 + L_1(\delta_1 - \hat\delta_1) \\
\dot{\hat\delta}_2 &= -H_{21}(q,\delta_1)(f_1(q,\dot q) + g_1(x_c)) \\
&\quad - H_{22}(q,\delta_1)(f_2(q,\dot q) + g_2(x_c) + K\hat\delta_1) \\
&\quad + H_{21}(q,\delta_1)u + (-H_{22}(q,\delta_1)K + L_2)(\delta_1 - \hat\delta_1) \quad (4.9)
\end{aligned}
$$

where $x_c^T = [q^T \ \dot q^T \ \delta_1^T \ \hat\delta_2^T]$ is the available state vector that may be used for control purposes and L_1 and L_2 are observer gain matrices to be selected. Defining the estimation errors $\tilde\delta_1 = \delta_1 - \hat\delta_1$, $\tilde\delta_2 = \delta_2 - \hat\delta_2$, the observer error dynamics can be obtained by subtracting (4.9) from (4.8), i.e.,

$$
\dot{\tilde\delta} = A_{\tilde\delta}\tilde\delta + b_{\tilde\delta}(x,\hat\delta_2) \quad (4.10)
$$

where $\tilde{\delta}^T = [\tilde{\delta_1}^T \ \tilde{\delta_2}^T]$ and

$$A_{\tilde{\delta}} = \begin{bmatrix} -L_1 & I \\ -L_2 & 0 \end{bmatrix}$$

$$b_{\tilde{\delta}}(x, \hat{\delta}_2) = \begin{bmatrix} 0 \\ H_{21}(-\frac{\partial g_1}{\partial \delta_2}\mid_{\hat{\delta}_2} \tilde{\delta}_2 + O_1(\tilde{\delta}_2{}^2)) + H_{22}(-\frac{\partial g_2}{\partial \delta_2}\mid_{\hat{\delta}_2} \tilde{\delta}_2 + O_2(\tilde{\delta}_2{}^2)) \end{bmatrix}.$$

$$(4.11)$$

The terms in (4.11) have been obtained by using the Taylor series expansions of g_1 and g_2 and noting that $\hat{\delta}_2 = \delta_2 - \tilde{\delta}_2$, i.e.,

$$g_1(x_c) = g_1(q, \dot{q}, \delta_1, \delta_2) - \frac{\partial g_1}{\partial \delta_2}\mid_{\hat{\delta}_2} \tilde{\delta}_2 + O_1(\tilde{\delta}_2{}^2)$$

$$g_2(x_c) = g_2(q, \dot{q}, \delta_1, \delta_2) - \frac{\partial g_2}{\partial \delta_2}\mid_{\hat{\delta}_2} \tilde{\delta}_2 + O_2(\tilde{\delta}_2{}^2). \quad (4.12)$$

It should be noted that since the components of δ_2 in g_1 and g_2 are at most of second order (centrifugal terms), the above Taylor series expansions terminate after the square terms. Moreover, considering a finite region Ω_o around the desired point $(x, \hat{\delta}_2)$, it follows that $\|b_{\tilde{\delta}}(x, \hat{\delta}_2)\| < k\|\tilde{\delta}\|$. If L_2 is of the same order of magnitude as H_{21} and H_{22}, the coefficient k is typically small. This follows by noting that $\frac{\partial g_1}{\partial \delta_2}\mid_{\hat{\delta}_2}$ and $\frac{\partial g_2}{\partial \delta_2}\mid_{\hat{\delta}_2}$ are $O(\delta\hat{\delta})$. Moreover, the matrix $A_{\tilde{\delta}}$ can be made Hurwitz by a proper choice of the gains L_1 and L_2, e.g., by selecting L_1 and L_2 to be any positive definite matrices. Thus, choosing a Lyapunov function candidate $V_o = \tilde{\delta}^T P_{\tilde{\delta}} Lyapunov! function \tilde{\delta}$, where $P_{\tilde{\delta}}$ is the solution of the Lyapunov equation

$$A_{\tilde{\delta}}^T P_{\tilde{\delta}} + P_{\tilde{\delta}} A_{\tilde{\delta}} = -Q_{\tilde{\delta}} \quad (4.13)$$

it follows that $\dot{V}_o \leq -(\lambda_{min}(Q_{\tilde{\delta}}) - 2\lambda_{max}(P_{\tilde{\delta}})k)\|\tilde{\delta}\|^2$. Hence, provided that $\lambda_{min}(Q_{\tilde{\delta}}) > 2k\lambda_{max}(P_{\tilde{\delta}})$ it follows that the error dynamics are locally asymptotically stable. Note that if parametric uncertainties are included, the error dynamics will converge to a residual set around the origin as shown in the sequel. In this case, assuming a bounded input vector u, it follows that $\|b_{\tilde{\delta}}\| < k\|\tilde{\delta}\| + k_u$, where k_u is an upper bound on all the terms due to parametric uncertainties. Then assuming that $\lambda_{min}(Q_{\tilde{\delta}}) > 2k\lambda_{max}(P_{\tilde{\delta}})$

$$\dot{V}_o \leq -(\lambda_{min}(Q_{\tilde{\delta}}) - 2\lambda_{max}(P_{\tilde{\delta}})k)\|\tilde{\delta}\|^2 + 2\lambda_{max}(P_{\tilde{\delta}})k_u\|\tilde{\delta}\|. \quad (4.14)$$

Furthermore, V_o can be written as

$$\lambda_{min}(P_{\tilde{\delta}})\|\tilde{\delta}\|^2 \leq V_o \leq \lambda_{max}(P_{\tilde{\delta}})\|\tilde{\delta}\|^2. \tag{4.15}$$

It then follows from (4.14) and (4.15) that for all $\tilde{\delta} \in \Omega_o$ we have

$$\dot{V}_o \leq -\gamma_1 V_o + \gamma_2 V_o^{1/2} \tag{4.16}$$

where

$$\gamma_1 = (\lambda_{max}(Q_{\tilde{\delta}}) - 2\lambda_{max}(P_{\tilde{\delta}}))/\lambda_{max}(P_{\tilde{\delta}}), \quad \gamma_2 = 2\lambda_{max}(P_{\tilde{\delta}})/\lambda_{min}(P_{\tilde{\delta}}).$$

Considering (4.16), we can conclude that if $V_o(0) > \gamma_2^2/\gamma_1^2$, then $\dot{V}_o < 0$. Thus the smallest residual set can be defined as $T_{res} = \{\tilde{\delta} \mid V_o \leq \gamma_2^2/\gamma_1^2\} \subset \Omega_o$ for which γ_2^2/γ_1^2 is minimum.

4.3.2.   Reduced–order Observer

It is possible to obtain a reduced–order observer using a similar technique as above. To this end, consider the following observation law

$$\begin{aligned}
\dot{\hat{\delta}}_2 &= -H_{21}(q, \delta_1)(f_1(q, \dot{q}) + g_1(x_c)) \\
&- H_{22}(q, \delta_1)(f_2(q, \dot{q}) + g_2(x_c) + K\delta_1) \\
&+ H_{21}(q, \delta_1)u + G(\hat{\delta}_2 - \delta_2)
\end{aligned} \tag{4.17}$$

where G is a Hurwitz gain matrix and x_c is as defined previously. The above equation reflects the fact that $\delta_2 = \dot{\delta}_1$ is not measurable. However, by taking $G\delta_2$ to the left–hand side of the equation and defining the auxiliary state variable $\hat{z} = \hat{\delta}_2 + G\delta_1$, it may be shown that

$$\begin{aligned}
\dot{\hat{z}} &= G\hat{z} - H_{21}(q, \delta_1)(f_1(q, \dot{q}) + g_1(x_c)) - H_{22}(q, \delta_1)(f_2(q, \dot{q}) \\
&+ g_2(x_c) + K\delta_1) + H_{21}(q, \delta_1)u - G^2\delta_1.
\end{aligned} \tag{4.18}$$

Similarly, defining an auxiliary state variable for the system dynamics as $z = \delta_2 + G\delta_1$, and subtracting the second equation in the resulting dynamics from (4.18), it follows that

$$\dot{e}_z = Ge_z + b_{\tilde{\delta}_2}(x, \hat{\delta}_2) \tag{4.19}$$

where $b_{\tilde{\delta}_2}(x, \hat{\delta}_2)$ is the second element of $b_{\tilde{\delta}}(x, \hat{\delta}_2)$ given by (4.11) and $e_z = z - \hat{z}$. Thus, if G is a Hurwitz matrix the error dynamics are stabilized locally. Note that (4.18) is now implementable. Thus, once $\hat{z}$ is obtained then $\hat{\delta}_2$ may be obtained from $\hat{\delta}_2 = \hat{z} - G\delta_1$. When parametric uncertainties are present a similar analysis as before can be developed to guarantee asymptotic stability of the error dynamics.

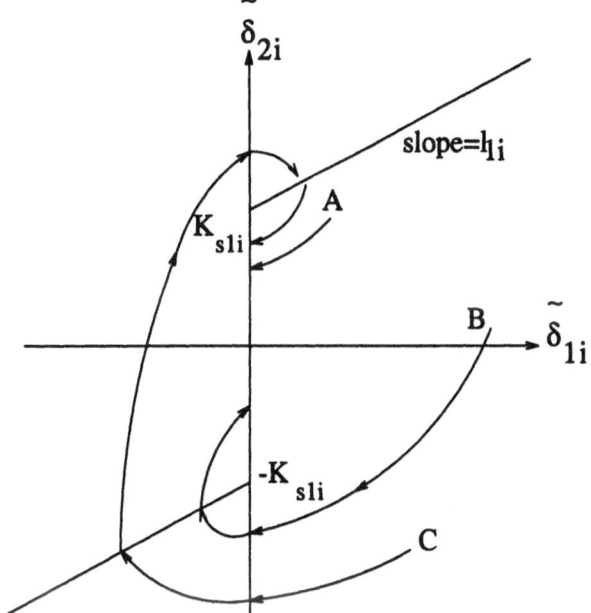

Figure 4.1. Phase–plane trajectories for the sliding observer.

4.3.3.   Sliding Observer

The sliding technique introduced in [61] and mentioned in the Section 4.1 may be incorporated in the full–order observer design established earlier. Towards this end, let L_1 and L_2 be positive definite diagonal matrices with diagonal elements l_{1i} and l_{2i}, $i = 1 \cdots m$, respectively. Then, by adding terms $K_{s1i}sgn(\tilde{\delta}_{1i})$ and $K_{s2i}sgn(\tilde{\delta}_{2i})$ ($sgn(.)$ is the signum function) to the observation laws (4.9), the resulting observer dynamics can be written as

$$\dot{\tilde{\delta}}_{1i} = -l_{1i}\tilde{\delta}_{1i} + \tilde{\delta}_{2i} - K_{s1i}sgn(\tilde{\delta}_{1i})$$
$$\dot{\tilde{\delta}}_{2i} = -l_{2i}\tilde{\delta}_{1i} - K_{s2i}sgn(\tilde{\delta}_{1i}) + \Delta f_i, \quad i = 1, \cdots, m \quad (4.20)$$

where Δf_i contains perturbation terms and modeling uncertainties. Then for each $\tilde{\delta}_{1i}$–$\tilde{\delta}_{2i}$, the sliding condition is satisfied in the region $\tilde{\delta}_{2i} \leq K_{s1i} + l_{1i}\tilde{\delta}_{1i}$, if $\tilde{\delta}_{1i} > 0$ and $\tilde{\delta}_{2i} \geq -K_{s1i} + l_{1i}\tilde{\delta}_{1i}$, if $\tilde{\delta}_{1i} < 0$. The dynamics on the *sliding patch* ($| \tilde{\delta}_{2i} | < K_{s1i}$) are derived from Filippov's solution concept [66], i.e., $\dot{\tilde{\delta}}_{2i} = -(K_{s2i}/K_{s1i})\tilde{\delta}_{2i} + \Delta f_i$. Now if K_{s2i} is selected such that $| \Delta f_i | < K_{s2i}$ the phase–plane trajectories are in the form given in Figure 4.1. Note that the slope l_{1i} affects regions of direct attraction to the sliding region.

In the rest of this chapter we will incorporate the observer strategies introduced above into the inversion based control law given by (4.6). The developments are carried out for the case of a full–order observer. However, a similar analysis can be done for the reduced–order observer.

4.4. Observer Based Inverse–dynamics Control

In this section we use the control law introduced in Section 4.2 except that $\dot{\delta}$ is replaced by $\hat{\dot{\delta}}$ given by one of the observer strategies discussed in Section 4.3. Since the redefined output velocity is not available we use its estimated value from $\dot{q} + \Psi\hat{\delta}_2$. Thus, the estimated redefined output velocity tracking error can be defined as $\hat{\dot{e}} = \dot{y}_r - (\dot{q} + \Psi\hat{\delta}_2)$ and the redefined output tracking error by $e = y_r - y$. In these relationships y_r and $\dot{y}_r$ are the reference trajectory and its velocity profile, respectively. Choosing $v = \ddot{y}_r + K_p e + K_d \hat{\dot{e}}$ with K_p, K_d determined by the output position error dynamics, yields

$$\dot{E} = A_E E + d_E(\alpha, x_c, t) \tag{4.21}$$

with

$$
\begin{aligned}
d_E(\alpha, x_c, t) &= -BK_\delta(q)\delta - BK_{\dot{\delta}}(q)\hat{\dot{\delta}} + O(\|K_d\|\tilde{\delta}) + O(\delta\dot{\delta}\tilde{\delta}) \\
E^T &= \begin{bmatrix} e^T & \dot{e}^T \end{bmatrix} \\
A_E &= \begin{bmatrix} 0 & I \\ -K_p & -K_d \end{bmatrix}.
\end{aligned} \tag{4.22}
$$

Moreover, (4.8) can be written in terms of u given by (4.6), that is

$$\dot{\Delta} = A_\Delta(q)\Delta + \begin{bmatrix} 0 \\ G_\Delta(x, \hat{\delta}, t) \end{bmatrix} \tag{4.23}$$

where

$$
\begin{aligned}
G_\Delta(x, \hat{\delta}, t) &= H_{21}B^{-1}(\ddot{y}_r + K_p e + K_d \dot{e}) - P(q)(f_2(q, \dot{q}) + g_2(x_c)) \\
&\quad + O(\delta) + O(\delta\tilde{\delta}) \\
A_\Delta(q) &= \begin{bmatrix} 0 & I \\ -P(q)K - H_{210}K_\delta(q) & -H_{210}K_{\dot{\delta}}(q) \end{bmatrix} \\
\Delta^T &= [\delta^T \quad \dot{\delta}^T]
\end{aligned} \tag{4.24}
$$

in which $H_{210} = H_{21}(q, 0)$ and

$$P(q) = [H_{220} - H_{210}(H_{110} + \Psi H_{210})^{-1}(H_{120} + \Psi H_{220})]. \quad (4.25)$$

Furthermore, matrices $K_{\delta}(q)$ and $K_{\dot{\delta}}(q)$ are selected such that $A_{\Delta}(q)$ is a Hurwitz matrix for the range in which q is varied. This can be guaranteed if the pair $(B_{\Delta_0}, A_{\Delta_0})$ is locally controllable on the domain of interest with

$$A_{\Delta_0}(q) = \begin{bmatrix} 0 & I \\ -P(q)K & 0 \end{bmatrix}$$

and

$$B_{\Delta_0}(q) = \begin{bmatrix} 0 \\ H_{210} \end{bmatrix}. \quad (4.26)$$

Noting that A_E, A_{Δ}, and $A_{\dot{\delta}}$ are Hurwitz matrices, it can then be shown (using a Lyapunov analysis) that the trajectories of the closed–loop system converge to a residual set with small tracking errors e and $\dot{e}$ and bounded δ and $\dot{\delta}$ provided that certain conditions are satisfied. The following theorem summarizes the above results

Theorem 4.1 *Let the control law (4.6) be applied to the original nonlinear system (4.1) with $B(q, \delta, \alpha)$ nonsingular in Ω_r (see (4.5)) and the pair $(B_{\Delta_0}, A_{\Delta_0})$ given by (4.26) controllable in Ω_r. Consider sets $\mathcal{R}$, $\mathcal{S}$ and $\mathcal{T}$ given by (A.37). Provided that the desired trajectories and their time derivatives (at least up to order 2) are continuous and bounded, it then follows that the trajectories of E, $\epsilon\Delta$ and $\tilde{\delta}$ (ϵ is a small scaling factor as discussed in Appendix A.3), starting from $\mathcal{S} - \mathcal{R}$ converge to a residual set $\mathcal{T}$ that can be made small by proper choice of controller parameters, if d_E and G_{Δ} given by (4.22) and (4.24), respectively satisfy certain norm conditions ((A.32) and (A.33)) in a bounded region Ω_i (given in Appendix A.3) of the full state space $(E, \Delta, \tilde{\delta})$.*

Proof: The proof of the above result is established in Appendix A.3.

Remark: Choice of Observer Gain Matrices

It is revealed from the analysis in Section 4.3. that to ensure asymptotic stability of the observer dynamics, the ratio

$$r = \lambda_{max}(P_{\tilde{\delta}})/\lambda_{min}(Q_{\tilde{\delta}})$$

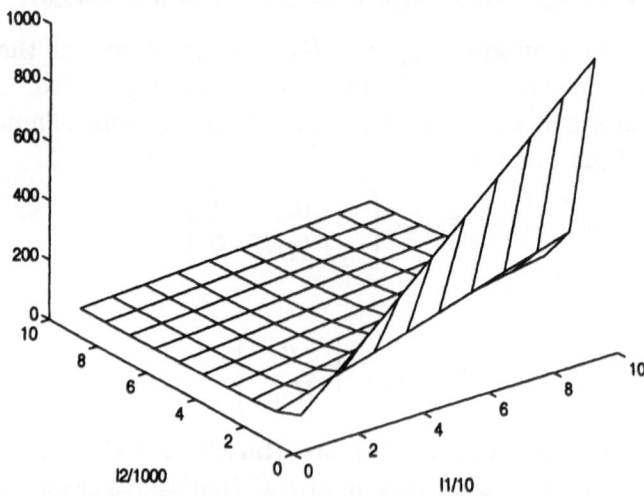

Figure 4.2. Plot of r versus observer gains.

has to be made sufficiently small. This will also reduce the size of the residual set. To this end, let us consider $Q_{\tilde{z}} = I$ (see e.g. [65]) and choose $L_1 = l_1 I$ and $L_2 = l_2 I$, where l_1 and l_2 are positive scalar parameters. A plot of r versus l_1 and l_2 is shown in Figure 4.2. It follows that observer gains have to be increased to achieve better observer robustness. For the sliding observer, these gains are infinite during sliding and are reduced outside the sliding region. Note, however that increasing the observer gains will increase sensitivity with respect to measurement noise. Thus, there is a limit as to how much the gains can be increased. Under certain noise conditions, Slotine *et al.* [61] have shown that sliding observers exhibit superior behavior when compared to Luenberger or Kalman filters.

4.5. Implementation of the Control Law

In this section the practical implementation of the control strategy discussed in this chapter is considered next. Figure 4.3 shows the schematic diagram of our experimental setup and Figure 4.4 shows the flexible-link system in different configurations. The flexible link is a stainless–steel $60cm \times 5cm \times 0.9mm$ rectangular bar with a $0.251kg$ payload attached to its end point. The mass of the bar is $0.216kg$

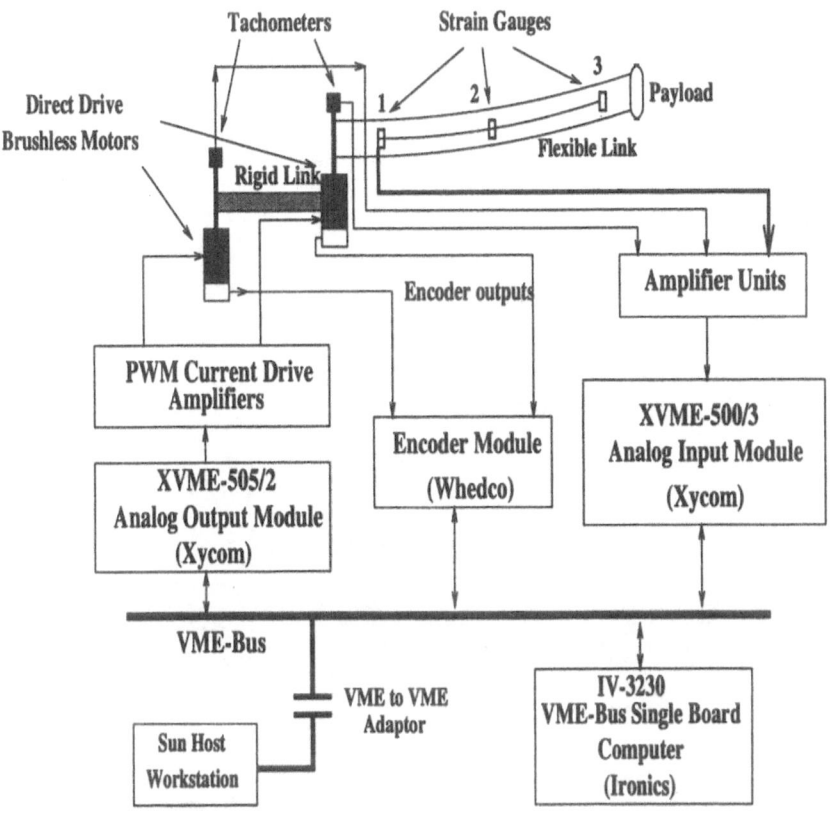

Figure 4.3. Experimental setup for the flexible–link robot.

that is comparable to its payload. The first link is a 20cm rigid aluminum bar. The two–link set up has significant nonlinear and non–minimum phase characteristics and its dynamics exhibit non-linearities that are similar to the case where both links are flexible. The first two flexible modes of this system when linearized around zero joint angles are 5.6 and 27.6 Hz. The roots corresponding to the linearized zero dynamics (when the tip position is taken as the output) are at $\pm j76.9$ and ± 16.0. This setup is essentially the same as one described in Chapter 2. The sensory equipment consists of three strain gauge bridges, two tachometers and two shaft–encoders that are used to measure the flexible modes of the link, joint rates, and joint positions, respectively. The signals from the strain gauge bridges and the tachometers are then amplified using low–drift amplifier stages and further passed through anti–aliasing filters. These signals are then fed into the *XVME-500/3* analog input module from

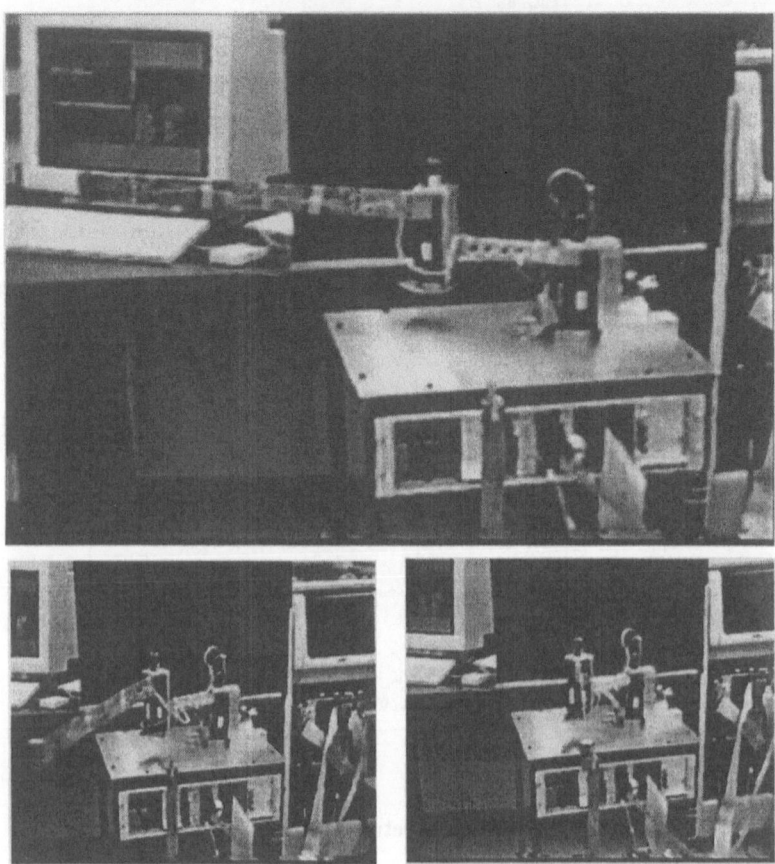

Figure 4.4. The flexible-link manipulator in different configurations.

Xycom. The actuators are *5113 Pittman* DC brushless servomotors which are direct driven by *503 Copley* PWM servo–drive amplifiers. The digital hardware has been selected based on the idea of a *reconfigurable* sensor–based control application as described in Chapter 2. The tip deflection is again constructed based on the measurements obtained from the strain gauges (see Chapter 2).

Figure 4.5 represents different software modules that were implemented using Chimera 3.1 based on the idea of reconfigurable subsystems. In this figure, five subsystems are indicated which exchange information through a shared-memory mechanism called global *State Variable Table* (SVT). Each subsystem, or module, is a separate task. The state-variable mechanism allows the frequency of each task to

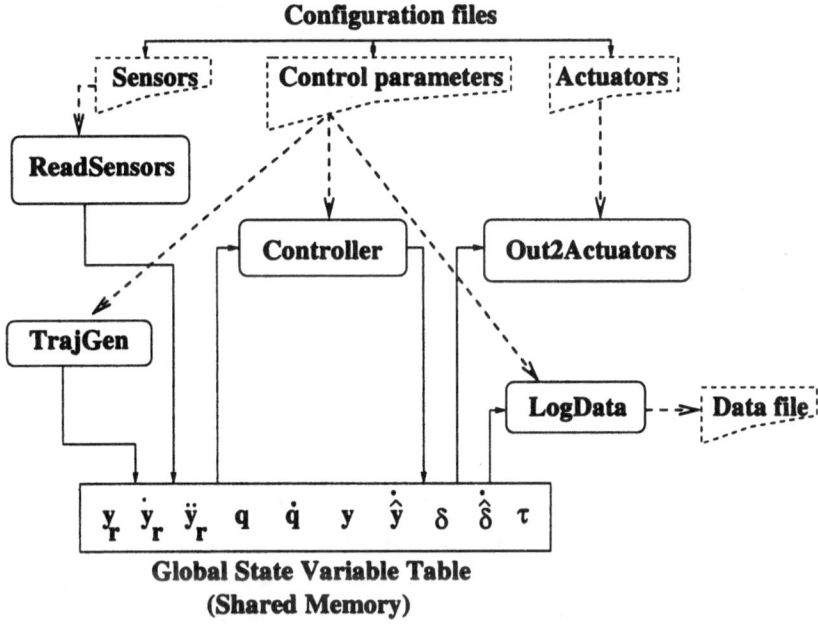

Figure 4.5. Control module software implementation under Chimera.

be different. For example the data logging task (*LogData*) in Figure 4.5 can be set to a value lower than other tasks which are running at a frequency of 350 Hz. The *ReadSensors* module obtains information from sensors and performs necessary convertions to update the shared memory with q, $\dot{q}$, δ. The conversion factors are loaded into memory by reading a configuration file which is performed once in the initialization phase. All other parameters such as final trajectory positions, controller parameters, data logging rates, and conversion factors for actuators needed by different modules are also loaded from disk files into memory during module initialization. The *TrajGen* module updates the SVT with desired tip-position trajectory data. The *Controller* module is a data producer and consumer task and updates the SVT with torques that have to be sent to the motors through the *Out2Actuators* module. Similarly *LogData* module is a consumer task that obtains the data during robot motion and writes it on a disk file for future analysis at the end of the motion profile. Data integrity is ensured by a providing a single locking mechanism for the shared-memory (SVT) [73].

The differential equations corresponding to a given observation strategy have to be solved numerically for a digital implementation.

The numerical algorithm has to be fast enough so that the results are computed and made available well before the end of each sampling period. The procedure adopted here is the *modified midpoint* method [74] which was also used in Chapter 2. The implemented algorithm took approximately 2 msec on the MC68030 Ironics processor board. Thus, a sampling frequency of 350 Hz was used. This rate was sufficient to allow computation of the control law as well as the data acquisition and trajectory generation tasks while maintaining closed–loop system stability.

Due to mechanical imprecisions in the physical construction of the manipulator the arm is not completely level in the horizontal plane. This can lead to considerable errors due to the fact that the magnitude of the required torque for control is specifically small at stopping points. The problem can be resolved to some extent by adding a compensating torque in the form of a Taylor series expansion of the static gravity torques up to the third power of joint angles. The Taylor series coefficients were thus estimated by measuring the balancing torques at several joint angles and using the least–squares algorithm.

4.5.1. Experimental Results

The control law for the two–link system was obtained based on the dynamic model given in Appendix C.2. The model was obtained by using the symbolic manipulation software *MAPLE* [35] but only *one* flexible mode was used in system modeling for control design. It would be reasonable to expect that the control performance based on a model with two flexible modes should result in better performance. However, it was found that the model based on a single mode gave better results. The explanation for this is that the mass matrix of the flexible–link system becomes more ill–conditioned as the number of flexible modes is increased. As a result, the inverted matrix becomes more sensitive to modeling errors and uncertainties, leading to poor or even unstable performance.

Figure 4.6 shows the condition number of the mass matrix as a function of time for a quintic trajectory tracking control of the second flexible arm shown in Figure 4.3. The problem can further be explained as follows. In practice the estimates $\hat{B}$ and $\hat{a}$ of B and a respectively in (4.6) are at our disposal. Also note that these terms are both affected by the inverse of the mass matrix. Thus taking

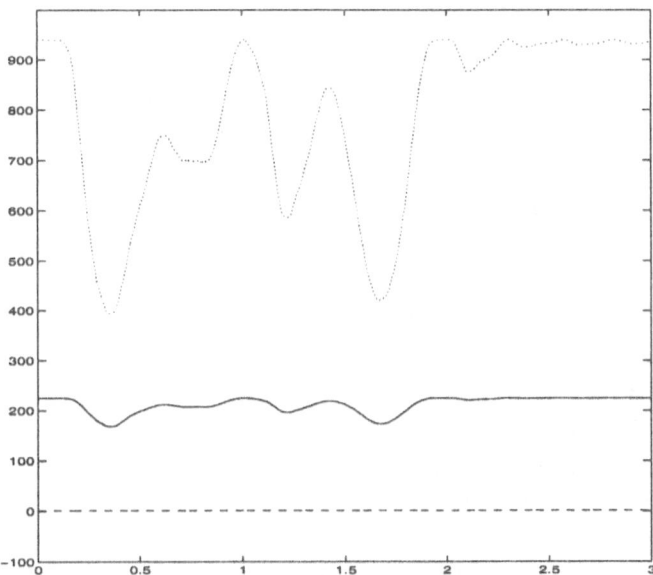

Figure 4.6. Condition number of the mass matrix vs time (s) for the flexible arm: Two mode model ($\cdots$), one mode model (—), rigid model (– – –).

$B^{-1} = \hat{B}^{-1} - \Delta B_I$ and $a = \hat{a} - \Delta a$, where ΔB_I and Δa are the error terms, the tracking error dynamics can be obtained from (4.7), i.e.,

$$
\begin{aligned}
\ddot{e} + K_d\dot{e} + K_p e & = \Delta B_I(v - a + \Delta a) + B^{-1}\Delta a \\
& + B(K_\delta(q)\delta + K_{\dot{\delta}}(q)\dot{\delta}).
\end{aligned}
\tag{4.27}
$$

Now, with a more ill–conditioned mass matrix the error terms, ΔB_I and Δa are likely to increase and destabilize the system. Thus there are two conflicting requirements in this regard: The ill–conditioning due to an increased number of modes and loosing control over higher flexural modes due to neglecting them in the control law. Both these factors can lead to instability. Therefore, the question of how many modes to take for satisfactory performance is difficult to quantify and should be decided in practice.

In all experiments, the *redefined* output of the second link was chosen to correspond to the angle from hub to the point at $0.8l_2$ (l_2 is the length of the second link). This location was obtained to be as near the tip position as possible while maintaining the non–minimum phase characteristic (as described in Chapter 3). The output of the first link is the joint angle itself. However, in all of the figures, the tip

output is reported instead of the re–defined output. Figure 4.7 shows
the results when a reduced–order observer is used to estimate rates
of change of the deflection modes and the control goal is to track a
3 second quintic polynomial trajectory with a 2 second tail of zero
velocity and acceleration. The observer gain was $G = -120$ and the
error dynamics gains were $K_p = 22.1I_{2\times2}$ and $K_d = 9.4I_{2\times2}$. K_δ and
$K_{\dot\delta}$ were zero since the small damping of the flexible–link was enough
to guarantee stability. Similarly, the results for the other observation
schemes are given in Figures 4.8 and 4.9 with $L_1 = 100I_{2\times2}$, $L_2 =$
$19000I_{2\times2}$, $K_p = 30.3I_{2\times2}$, $K_d = 11.0I_{2\times2}$ for the full–order observer
and $l_{11} = 1$, $l_{21} = 600$, $K_{s11} = 5$, $K_{s21} = 600$, $K_p = 22.1I_{2\times2}$,
$K_d = 9.4I_{2\times2}$ for the sliding–mode observer. K_δ and $K_{\dot\delta}$ were 0 and
$0.2B^{-1}$, respectively for the full–order observer but were chosen as 0
and $0.1B^{-1}$ for the case with the sliding–mode observer. The choice
of K_δ and $K_{\dot\delta}$ in this way is first to ensure that $A_\Delta(q)$ given by 4.24
is Hurwitz and second to ensure that these terms are not too large
in 4.6 as this will destabilize the closed–loop system.

It is observed from all the figures that even after the 3 second
quintic trajectory is over there is some control activity. This is due
to the gravity effects that bend the link at the stopping point and
result from the compensation provided by the controller.

For the reduced–order observer, the closed–loop system became
unstable when G was increased. Similar results were obtained for
the other two observers. This can be attributed to the increased
noise sensitivity for higher gains. In the sliding–mode observer the
ratio K_{s21}/K_{s11} had to be chosen sufficiently large for system stability. This is mainly due to the fact that on the sliding patch the
disturbance terms have to be rejected. Moreover, the operation of
the controller was associated with control chattering. Although the
bandwidths of the actuators were small enough not to pass the high
frequency chattering signals, the operation of the closed–loop system
was more oscillatory than the other observers. The problem was resolved by replacing the $sgn(.)$ with a saturation nonlinearity of the
following form

$$sat(x, \varepsilon) = \begin{cases} 1 & x > \varepsilon \\ x/\varepsilon & -\varepsilon \leq x \leq \varepsilon \\ -1 & x < -\varepsilon \end{cases}$$

with $\varepsilon = 0.05$.

Overall, the full–order observer was found to result in a better closed–loop performance with little tuning of the control and observer gains.

Figures 4.10 and 4.11 show the experimental results with the conventional computed torque PD control when the effects of flexibility are neglected in the plant model. This is done to assess the performance improvement of the proposed scheme over a conventional method. Figure 4.10 corresponds to the case when the same K_p and K_p are used as before. The control input was disconnected due to instability of the closed–loop system. On the other hand, reducing the gains to $K_p = 1$ and $K_d = 2$ results in large tracking errors and increased sensitivity of the closed–loop system to uncertainties as shown in Figure 4.11.

Due to several factors the tracking errors are not as small as those predicted by the simulations which were run in the absence of modeling imperfections, actuator dynamics, discretization effects, sensor noise, and computational delay of the control law.

4.6. Conclusion

A control strategy based on nonlinear inversion was proposed for a class of structurally flexible robot manipulators and experimentally tested on a two–arm flexible setup. Three observation strategies were also introduced to estimate the rates of change of flexible modes that are not conveniently or economically accessible. The observation strategies are applicable to the general case of flexible robots and tend to cancel certain nonlinear terms that are present in the dynamic equations so that linear observation error dynamics are obtained. The flexible robot was also tested with a conventional control method based on rigid–link inverse dynamics PD control. The proposed scheme shows promising results for stable and small tip position tracking error performance.

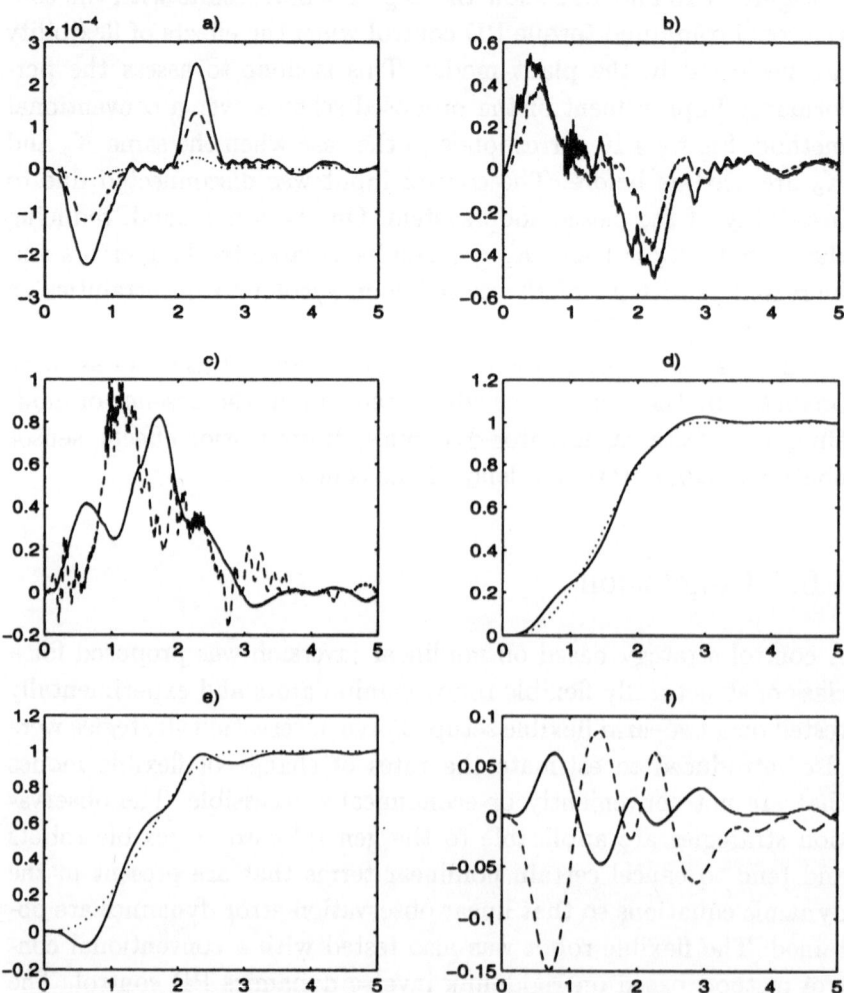

Figure 4.7. Experimental results using the reduced–order observer ($x-axis$: time (s)).   a) Strain measurements at points 1 (—), 2 (--), and 3 ($\cdots$) (m/m)   b) Actuator inputs (Nm): 1 (—), 2 (--)   c) Joint velocities (rad/s): 1 (—), 2 (--)   d) First joint angle (—) and desired trajectory ($\cdots$) (rad)   e) Second link tip angle (—) and desired trajectory ($\cdots$) (rad)   f) Trajectory tracking errors (rad): 1 (—), 2 (--).

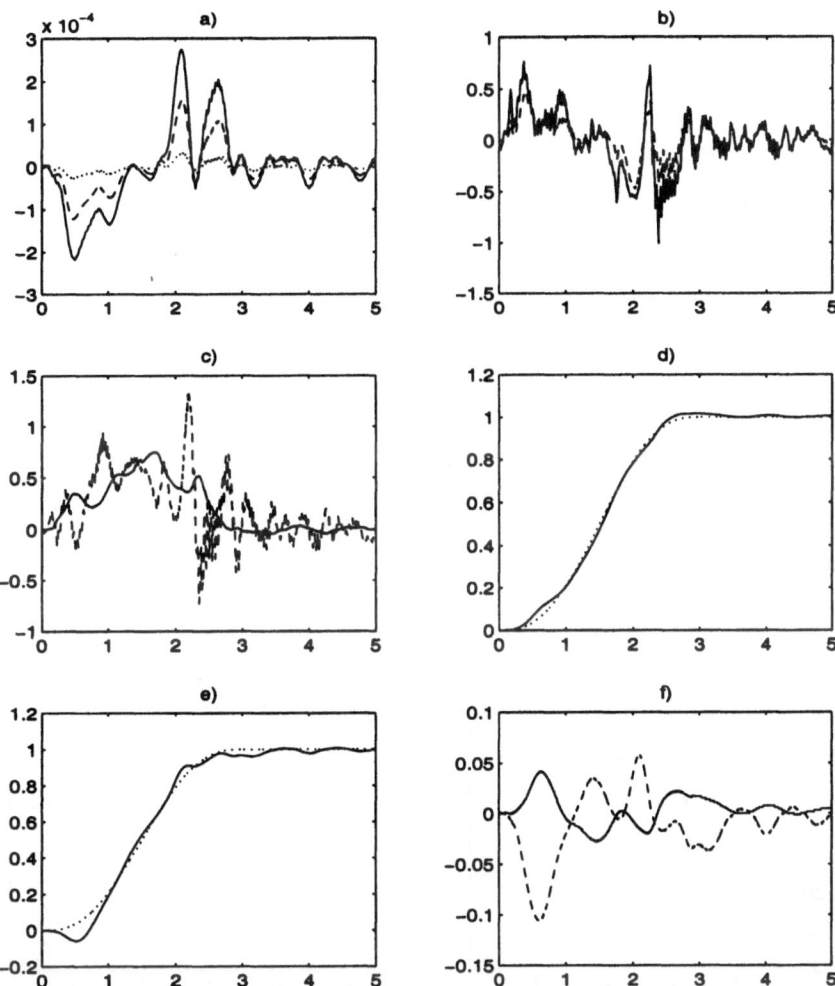

Figure 4.8. Experimental results using the full–order observer ($x - axis$: time (s)). a) Strain measurements at points 1 (—), 2 (--), and 3 (···) (m/m) b) Actuator inputs (Nm): 1 (—), 2 (--) c) Joint velocities (rad/s): 1 (—), 2 (--) d) First joint angle (—) and desired trajectory (···) (rad) e) Second link tip angle (—) and desired trajectory (···) (rad) f) Trajectory tracking errors (rad): 1 (—), 2 (--).

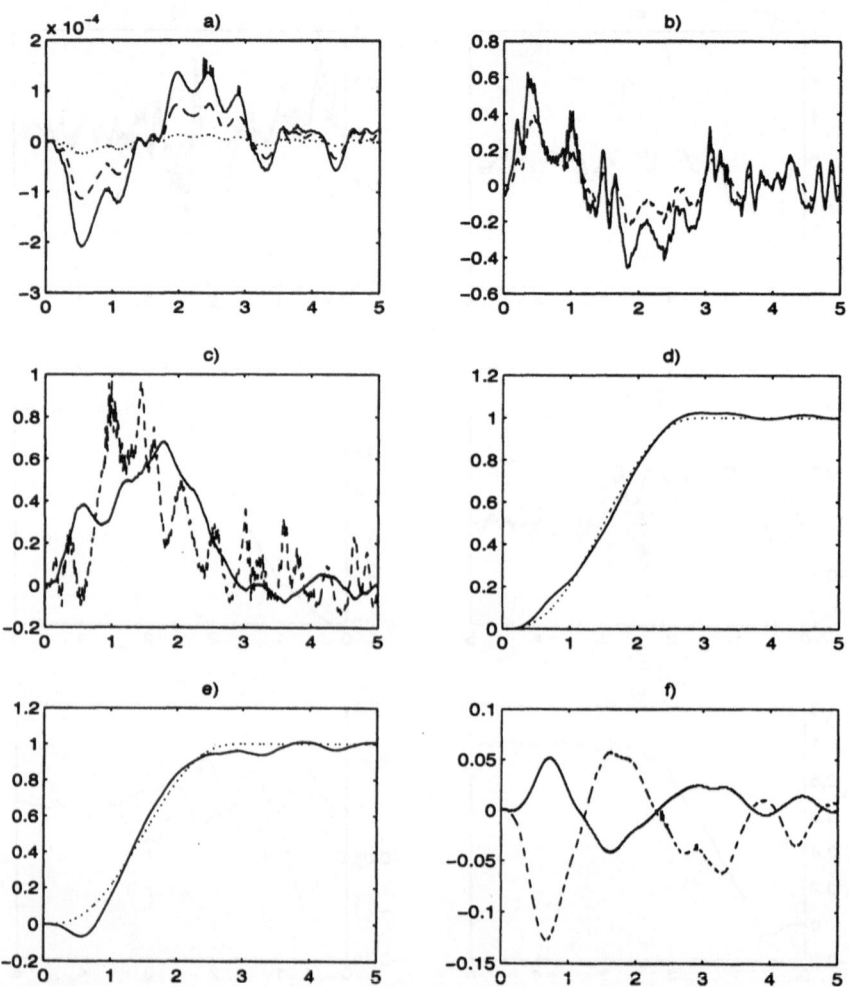

Figure 4.9. Experimental results using the sliding–mode observer ($x - axis$: time (s)).   a) Strain measurements at points 1 (—), 2 (−−), and 3 ($\cdots$) (m/m)   b) Actuator inputs (Nm): 1 (—), 2 (−−)   c) Joint velocities (rad/s): 1 (—), 2 (−−) d) First joint angle (—) and desired trajectory ($\cdots$) (rad)   e) Second link tip angle (—) and desired trajectory ($\cdots$) (rad)   f) Trajectory tracking errors (rad): 1 (—), 2 (−−).

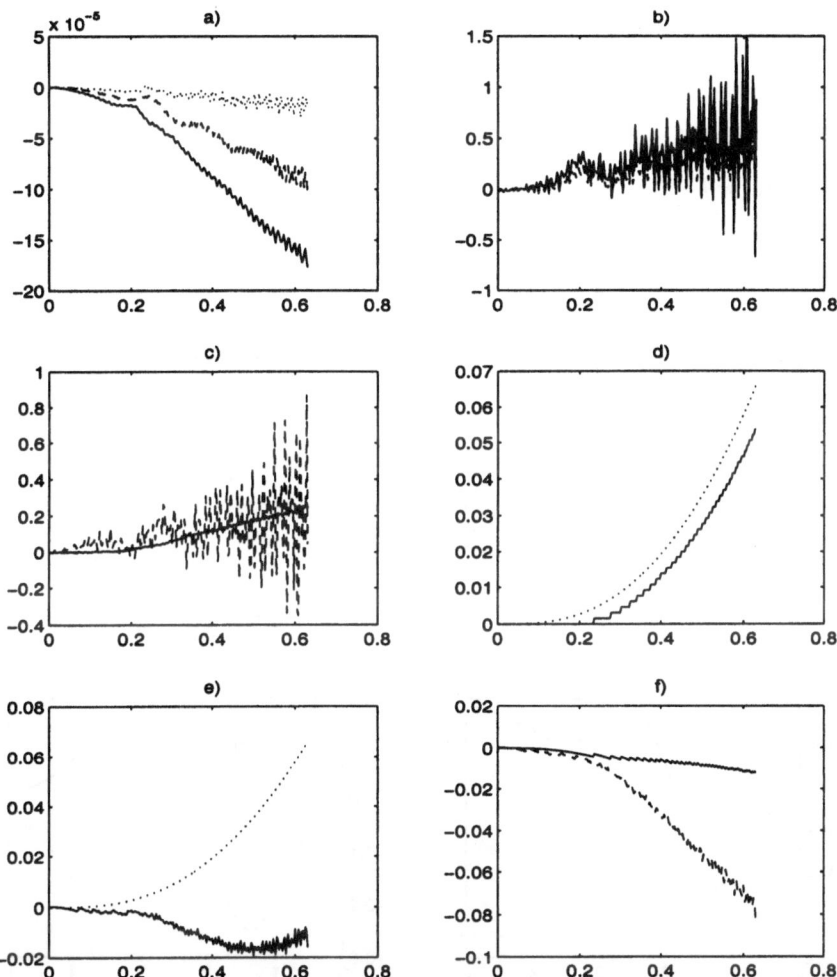

Figure 4.10. Unstable experimental results using conventional inverse dynamics PD controller ($x - axis$: time (s)).   a) Strain measurements at points 1 (—), 2 (– –), and 3 ($\cdots$) (m/m)   b) Actuator inputs (Nm): 1 (—), 2 (– –)   c) Joint velocities (rad/s): 1 (—), 2 (– –)   d) First joint angle (—) and desired trajectory ($\cdots$) (rad)   e) Second link tip angle (—) and desired trajectory ($\cdots$) (rad)   f) Trajectory tracking errors (rad): 1 (—), 2 (– –).

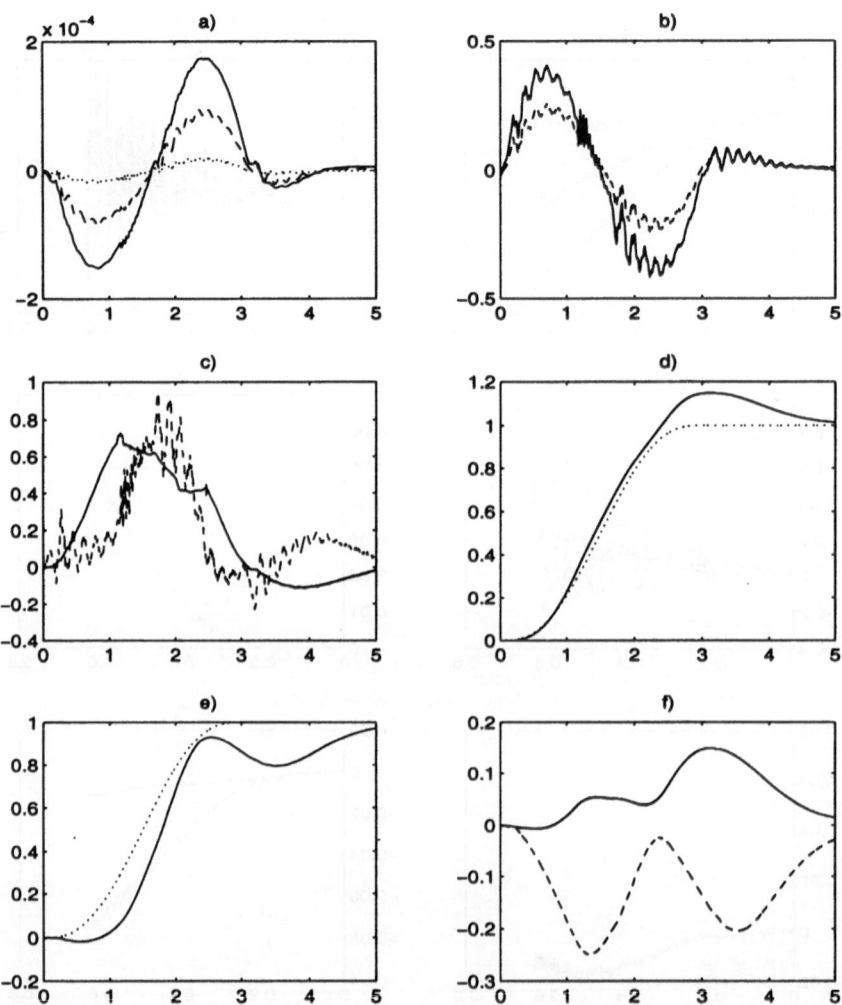

Figure 4.11. Experimental results using conventional inverse dynamics PD controller with reduced gains ($x-axis$: time (s)). a) Strain measurements at points 1 (—), 2 (−−), and 3 (···) (m/m)   b) Actuator inputs (Nm): 1 (—), 2 (−−) c) Joint velocities (rad/s): 1 (—), 2 (−−)   d) First joint angle (—) and desired trajectory (···) (rad)   e) Second link tip angle (—) and desired trajectory (···) (rad)   f) Trajectory tracking errors (rad): 1 (—), 2 (−−).

5.

Inverse Dynamics Sliding Control

In this chapter, we modify the control strategy based on dynamic inversion discussed in the last two chapters such that a more robust performance is achieved in the presence of considerable parametric uncertainties. Motivated by the concept of a *sliding surface* in variable structure control (*VSC*) [75], a robustifying term is developed to drive the nonlinear plant's error dynamics onto a sliding surface. On this surface, the error dynamics are then independent of parametric uncertainties. In order to avoid over–excitation of higher frequency flexural modes due to control chattering, the discontinuous functions normally used in classical *VSC* are replaced by saturation nonlinearities at the outset. This also facilitates analysis by standard Lyapunov techniques. The controller performance is demonstrated by simulation on a two–link flexible manipulator with considerable amount of parametric uncertainty.

5.1. Introduction

Because of the infinite dimensional nature of the dynamics of flexible–link manipulators and in order to obtain less complicated models for control design, approximations are usually made in the modeling phase. These approximations may be the source of parametric uncertainties which may in turn lead to poor or unstable performance. Payload variations can also significantly change the dynamic equations hence causing similar problems. One way to deal with this issue is employing concepts from variable structure control [75], [76] which features excellent robustness properties in the face of parametric uncertainties. This has been successfully applied for trajectory tracking control of rigid robot manipulators [77]. In the case of flexible–link manipulators, one has to deal with the non–minimum characteristic of the plant and to ensure that the control law is smooth enough not to over excite higher flexural modes of the plant.

Control design based on sliding surfaces is considered in this chapter. By defining a desired sliding surface and partitioning the control input, a control law is developed to ensure attractiveness of the surface, boundedness of the flexible modes, and tracking of the re–defined outputs. For simplicity we assume that all the states are available for control. Obviously the observer structure introduced in Chapter 4 may be employed to obtain flexural rates if desired.

A similar closed–loop stability analysis is performed and conditions for achieving stable closed–loop behavior are derived. The control scheme is further tested by a numerical simulation for a two–link flexible manipulator.

5.2.   Control Based on Input–output Linearization

Let us consider the control strategy discussed in Section 4.2. Assuming that the nominal vector $a_n(x)$ and matrix $B_n(\alpha, q, \delta)$, representing $a(\alpha, x)$ and $B(\alpha, q, \delta)$ respectively, are at our disposal, (4.3) can be written as

$$\ddot{y} = a_n(x) + B_n(q, \delta)u + \eta(x, u) \qquad (5.1)$$

where $\eta(x, u) = \Delta a(x) + \Delta B(q, \delta)u$ and $\Delta a(x) = a(\alpha, x) - a_n(x)$, $\Delta B(q, \delta) = B(q, \delta) - B_n(q, \delta)$. Let us define a desired trajectory profile y_r, $\dot{y}_r$, and $\ddot{y}_r$, the tracking errors $e = y_r - y$ and $\dot{e} = \dot{y}_r - \dot{y}$, and an additional state $e_I = \int_0^t e \, dt$. Then in the $3n$–dimensional space of $(e_I, e, \dot{e})$ define a linear n–dimensional switching surface

$$\sigma = K_p e_I + K_d e + \dot{e} \qquad (5.2)$$

where K_p and K_d are positive–definite matrices that, as will be seen later, determine the error dynamics in the sliding mode. Our goal is to design a control law such that two conditions are satisfied: First, on the sliding (switching) surface the dynamics are independent of the uncertainties, and second, the surface is an attractive manifold. As discussed in [80] in order to obtain smooth action let us define a boundary layer around the switching surface σ and measure the distance of each point to this surface by

$$s_\sigma = \sigma - \Phi sat(\frac{\sigma}{\Phi}). \tag{5.3}$$

Then inside the boundary layer, s_σ and $\dot{s}_\sigma$ are zero, while outside the boundary layer we have

$$\dot{s}_\sigma = \ddot{e} + K_d \dot{e} + K_p e. \tag{5.4}$$

Substituting $\ddot{y}$ from (5.1) in (5.4) yields

$$\dot{s}_\sigma = \ddot{y}_r + K_d \dot{e} + K_p e - a_n(x) - B_n(q, \delta)u - \eta(x, u). \tag{5.5}$$

Let us partition the control input as

$$u = u_n + u_u + u_\Delta \tag{5.6}$$

where u_n is the nominal control, u_u is to provide more robustness to uncertainties, and $u_\Delta = K_\delta(q)\delta + K_{\dot{\delta}}\dot{\delta}$, as will be seen later, guarantees boundedness of δ and $\dot{\delta}$.

Assuming that the nominal $B_n(q, \delta)$ is nonsingular on the domain of interest, let us take

$$u_n = B_n^{-1}(q, \delta)(-a_n(x) + \ddot{y}_r + K_d \dot{e} + K_p e). \tag{5.7}$$

Then substitution of (5.7) in (5.5) yields

$$\dot{s}_\sigma = -B_n u_u - \eta(x, u). \tag{5.8}$$

Thus to make s_σ an attractive manifold in the presence of parametric uncertainties let us choose

$$u_u = -B_n^{-1} K sat(\Gamma s_\sigma) \tag{5.9}$$

where $K = diag(k_1, \cdots, k_n)$, $\Gamma = diag(\gamma_1, \cdots, \gamma_n)$ and $sat(.)$ is the saturation vector function. Substituting (5.9) in (5.8) yields

$$\dot{s}_\sigma = K sat(\Gamma s_\sigma) - \eta(x, u). \tag{5.10}$$

Note that u is now a function of x, s_σ, t, or equivalently a function of e, $\dot{e}$, δ, $\dot{\delta}$, s_σ, t. Consider a bounded reference trajectory and e, $\dot{e}$, δ, and $\dot{\delta}$ inside a closed bounded set $\Omega^* \subset R^{2n+2m}$ then

$$\dot{s}_\sigma = K sat(\Gamma s_\sigma) + \Delta B^* K sat(\Gamma s_\sigma) + \eta^*(e, \dot{e}, \delta, \dot{\delta}) \qquad (5.11)$$

where $\Delta B^* = \Delta B B_n^{-1}$ and

$$\begin{aligned} \eta^*(e, \dot{e}, \delta, \dot{\delta}) &= -\Delta a(x) - \Delta B^*(-a_n + \ddot{y}_r + K_d \dot{e} + K_p e) \\ &\quad - \Delta B(K_\delta(q)\delta + K_{\dot{\delta}}\dot{\delta}). \end{aligned} \qquad (5.12)$$

Representing the elements of ΔB^* by ΔB_{ij}^*, (5.11) can be written as

$$\begin{aligned} \dot{s}_{\sigma_i} &= (1 + \Delta B_{ii}^*) k_i sat(\gamma_i s_{\sigma_i}) + \sum_{j=1, j\neq i}^{n} \Delta B_{ij}^* k_j sat(\gamma_j s_{\sigma_j}) + \eta_i^*, \\ i &= 1, \cdots, n. \end{aligned} \qquad (5.13)$$

Now for all $(e, \dot{e}, \delta, \dot{\delta}) \in \Omega^*$ assume that $\mid \Delta B_{ij}^* \mid \leq b_{ij}$ and $\eta_i^* \leq c_i$. Hence for all $s_{\sigma_i} \in R$ the sliding conditions can be individually ensured [66] if

$$s_{\sigma_i} \dot{s}_{\sigma_i} \leq -\beta_i \mid s_{\sigma_i} \mid, \quad \beta_i > 0, \quad i = 1, \cdots, n. \qquad (5.14)$$

Noting that $s_{\sigma_i} sat(\gamma_i s_{\sigma_i}) \leq \mid s_{\sigma_i} \mid$ it follows from (5.13) that (5.14) can be ensured under worst case conditions if k_i's are such that

$$(1 - b_{ii})k_i - \sum_{j=1, j\neq i}^{n} b_{ij}k_j - c_i \geq \beta_i. \qquad (5.15)$$

Considering the equality sign in (5.15) we have the matrix equation $(I - B^*)k^* = d$ where B^* is the matrix with elements b_{ij} and d is the vector with elements $c_i + \beta_i > 0$. Following Frobenius–Perron theorem (see e.g. [66]) a unique solution for k^* can be guaranteed with all its elements positive if the largest real eigenvalue of B^* is less than 1. In particular, if b_{ij}'s are chosen such that B^* is symmetric, the eigenvalue condition translates into the norm condition $\|B^*\|_2 < 1$. Moreover, if such a solution exists for k^*, choosing the elements of k larger than k^* will result in larger β_i's, hence speeding up attraction when the trajectories are outside $s_{\sigma_i} = 0$.

5.2.1. Stability of the Closed–loop System

To prepare for a stability analysis of the closed–loop system with the aforementioned control laws, let us first substitute (5.6) into (5.1) by utilizing (5.7) and (5.9). Then arranging the expressions in terms of e and $\dot{e}$ yields

$$\dot{E} = A_E E + d_E(E, \Delta, s_\sigma, t) \tag{5.16}$$

where $E^T = [e^T \quad \dot{e}^T]$, $\Delta^T = [\delta^T \quad \dot{\delta}^T]$, $d_E(.) = [0^T \quad D_E^T(.)]^T$ and

$$
\begin{aligned}
d_E(E, \Delta, s_\sigma, t) = \ & -(I + \Delta B) K sat(\Gamma s_\sigma) + (B_n + \Delta B)(K_\delta(q)\delta \\
& + \ K_{\dot{\delta}}\dot{\delta}) \\
& + \ \Delta B B_n^{-1}(-a_n + \ddot{y}_r + K_p e + K_d \dot{e}) \tag{5.17}
\end{aligned}
$$

and A_E is a Hurwitz matrix given by

$$A_E = \begin{bmatrix} 0 & I \\ -K_p & -K_d \end{bmatrix}. \tag{5.18}$$

For the part of dynamics due to the flexible modes, starting from (4.3) and substituting u will lead after some manipulations to

$$
\begin{aligned}
\ddot{\delta} = \ & -[H_{22} - H_{21}B^{-1}(H_{21} + \Psi H_{22}](f_2 + g_2 + K\delta) \\
& + \ H_{21}(B^{-1}a - B_n^{-1}a_n) \\
& + \ H_{21}B_n^{-1}(\ddot{y}_r + K_p e + K_d \dot{e} - K sat(\Gamma s_\sigma)) \\
& + \ H_{21}(K_\delta(q)\delta + K_{\dot{\delta}}(q)\dot{\delta}). \tag{5.19}
\end{aligned}
$$

Equation (5.19) is strongly linear in terms of δ and $\dot{\delta}$. Thus expanding the terms around $\delta = 0$ and $\dot{\delta} = 0$ results in

$$\dot{\Delta} = A_\Delta(q)\Delta + g_\Delta(E, \Delta, s_\sigma, t) \tag{5.20}$$

where $g_\Delta = [0^T \quad G_\Delta^T]^T$ and

$$
\begin{aligned}
A_\Delta(q) &= \begin{bmatrix} 0 & I \\ -P(q)K - H_{210}K_\delta(q) & -H_{210}K_{\dot{\delta}}(q) \end{bmatrix} \\
P(q) &= [H_{220} - H_{210}(H_{110} + \Psi H_{210})^{-1}(H_{120} + \Psi H_{220})] \tag{5.21}
\end{aligned}
$$

where $H_{ij0} = H_{ij}(q, \delta = 0)$ $(i, j = 1, 2)$. Moreover G_Δ is given by

$$
\begin{aligned}
G_\Delta(E, \Delta, s_\sigma, t) = \ & H_{21}B_n^{-1}(\ddot{y}_r + K_p e + K_d \dot{e} - K sat(\Gamma s_\sigma)) \\
& + \ H_{21}(B^{-1}a - B_n-1 a_n) + O(\delta, q, \dot{q}). \tag{5.22}
\end{aligned}
$$

The reason for writing this portion of the dynamics in the form just described is that (5.20) can be shown to be related to the zero–dynamics of the system with re–defined outputs which is made stable by an appropriate choice of the outputs and matrices $K_\delta(q)$ and $K_{\dot\delta}(q)$ [69].

The stability proof can then be achieved by considering the composite Lyapunov function candidate

$$V = E^T P_E E + \epsilon_\Delta^2 \Delta^T P_\Delta(q)\Delta + 0.5 s_\sigma^T s_\sigma \tag{5.23}$$

and finding the derivative of V along the trajectories of the system. This analysis is further detailed in Appendix A.4 and results in the following theorem

Theorem 5.1 *Let the control law (5.6) be applied to the original nonlinear system (4.1) with $B_n(q,\delta)$ nonsingular and let the flexible dynamics given by (5.20) be controlled by choice of $K_\delta(q)$ and $K_{\dot\delta}(q)$ on the domain of interest. Consider sets $\mathcal{R}$, $\mathcal{S}$ and $\mathcal{T}$ given by (A.48). Then assuming that the desired trajectories and their time derivatives (at least up to order 2) are continuous and bounded, it follows that the trajectories of E, $\epsilon_\Delta \Delta$ and s_σ (ϵ_Δ is a small scaling factor as discussed in Appendix A.4), starting from $\mathcal{S} - \mathcal{R}$ converge to a residual set $\mathcal{T}$ that can be made small by proper choice of controller parameters, provided that d_E and g_Δ given in (5.16), (5.20) satisfy certain norm conditions ((A.44), (A.45)) in a bounded region Ω_i (given in Appendix A.4) of the state space of (E, Δ, s_σ) containing the origin.*

Proof: The proof of the above result is established in Appendix A.4.

5.2.2. Numerical Simulations

In this section the controller performance is tested using a numerical simulation for the model of the two–link experimental setup that was discussed in Section 4.5 (Figure 4.3).

The nominal model used in the numerical simulation is based on the case where no payload is attached at the end point while the plant model is based on the full payload of $251kg$. This results in a significant difference between the parameter values of these models (of order of 2–6 times). Figure 5.1 shows the simulation results for

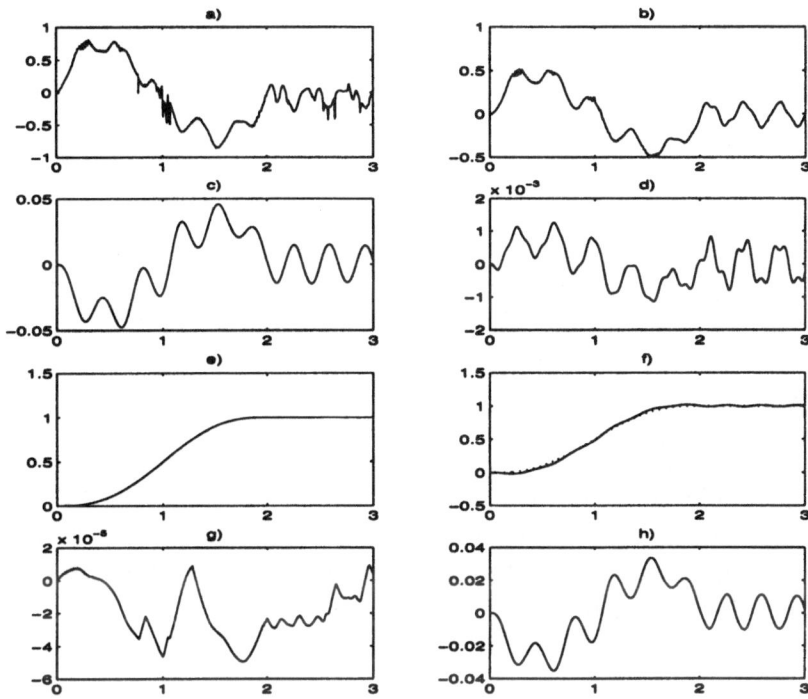

Figure 5.1. Simulation results for a flexible two–link manipulator using the proposed controller (Horizontal axis: time (s)):   a) First joint torque input (Nm)   b) Second joint torque input (Nm)   c) First flexible mode (m)   d) Second flexible mode (m)   e) First joint angle (—) and reference trajectory (···) (rad) f) Tip position of second link (—) and reference trajectory (···) (rad) g) Tracking error of the first link joint position (rad)   h) Tip position tracking error of the second link (rad).

$K = 6000$, $\Gamma = 2I$, $K_p = 1$, $K_d = 2$, $\Phi = 1.7 \times 10^{-4}$, $K_\delta = 0$ and $K_{\hat{\delta}} = 1$. The output of the first link was the joint angle while for the second link the output was chosen as the pseudo–angle corresponding to the point at $0.80L_2$ (L_2 is the length of the second link). It is observed that the tracking errors are small. Several other simulations were also performed by changing the gain parameter values and similar results were obtained. To compare the performance of the sliding controller introduced in this chapter with that of the control strategy in Chapter 3 similar simulations were performed. The closed–loop system consisted of a control law based on a plant model with zero payload while the actual plant model was based on a payload of $0.251kg$. The closed–loop system was unstable despite various changes made in controller gains. This result further demonstrates that the sliding controller is more robust than the controller introduced in Chapter 3 when parametric uncertainties are significant.

5.3.   Conclusion

The decoupling control strategy discussed in Chapters 3 was modified by using the concept of sliding surfaces so that more robustness to parametric uncertainties is achieved. Two main factors contribute to the robust performance of the closed–loop system in this case: The input–output linearization that tends to approximately decouple the dynamics, and the sliding control component that causes the off–manifold trajectories to be attracted to the surface. Full state measurements were necessary in the developed scheme; however, in a practical situation $\dot{\delta}$ can be estimated by using a nonlinear observer as described in Chapter 4.

6.

Optimum Structure Design for Control

In this chapter, an optimum structure design methodology is proposed for improving the dynamic behavior of structurally flexible manipulators. In this regard, a multi–objective optimization index is introduced for improving the dynamic behavior of an actuated structurally flexible arm through geometric shape design. The improvement in the dynamic behavior is achieved by defining a *vector* optimization cost function. This cost function is associated with the frequency of the lowest system zero when joint angles are taken as outputs, the gain sensitivity of open-loop system modes, and the gain sensitivity of the closed–loop system modes at its transmission zeros. Moreover, a general relationship is obtained between the system pole and zero locations. In particular, it is shown that the magnitude of each system zero is smaller than a corresponding flexural mode and thus the smallest zero always occurs before the smallest pole and should therefore be considered in structural shape design. The design method is applied for finding an optimized structural shape for a single–link flexible arm. The optimized link is shown to yield superior robustness and performance characteristics in a closed–loop system when compared to the non–optimized uniform link.

6.1. Introduction

Improving the plant characteristics for achieving a well–behaved system for the purpose of control design has been pursued in systems and control engineering for many years now. As an example, for an aircraft to be open–loop stable, the center of mass has to be ahead of the center of pressure. Thus, one desirable aspect of aircraft design is to guarantee that this condition is satisfied. The same philosophy can also be applied to flexible structure robots as described in [82], [21], [22]. The shape design process to achieve characteristics such as low mass and moments of inertia and high natural frequencies generally requires solution of an optimization problem. Although increasing the structural natural frequencies, as described in the earlier works, will help to improve system properties, it does not necessarily ensure that the plant behavior is improved in terms of control robustness. In this regard, an optimization procedure was introduced in [71] that incorporates an intuitive measure called *modal accessibility* [100] into the design process. In control systems theory, controllability and observability matrices provide basically binary decisions about whether the system is controllable or observable. This information cannot easily be incorporated into a continuous optimization problem and does not provide information on how close the system is to being unobservable or uncontrollable. However, controllability and observability grammians provide such information in a continuous manner although they are more difficult to obtain and utilize in practice. One objective of this chapter is to obtain measures that are related to these properties for flexible-structure systems such as flexible robot arms and then incorporate them in designing arm shapes for improved performance. In this regard, a measure denoted as *modal accessibility* has been introduced in [100] and [101], defined via the norms of the rows of input matrix for the diagonalized modal form (A, B) based on [102]. However, this measure is somewhat arbitrary and of limited use in determining the level of controllability of a mode in that it does not in general give an indication of the *distance* to uncontrollability of the mode.

In [102] controllability and observability measures are given for general linear systems by defining appropriate Frobenius norms of system matrices. In the present work, we obtain measures that are related to these system properties and utilize them in an optimization index to improve the dynamic behavior of a flexible arm system.

Towards this end, eigenvalue sensitivity measures associated with the flexural part of the manipulator dynamics are obtained. In flexible–link manipulators, these dynamics usually impose serious constraints on the closed–loop system performance and pose robustness limitations. It is shown that these sensitivities are related to the controllability and observability properties of the system. A high modal gain sensitivity reflects the fact that the modes can be easily moved to the left half–plane by providing proper feedback. Nevertheless, high sensitivities are not always desirable. In the case of output feedback, the closed–loop eigenvalues approach transmission zeros as the controller gains are increased. For a flexible–link manipulator, some of the transmission zeros are at critical locations, namely in the right half–plane if the tip deflections are taken as outputs, and on the $j\omega$–axis if the joint positions are taken as outputs. In either case, poor behavior will result when the gains are increased. In a joint feedback control strategy, having small gain sensitivity near the zero locations is desirable to ensure that the closed–loop system modes remain in *safe* locations in the left half–plane with proper feedback gains. On the other hand, having highly controllable open–loop modes will ensure robustness since smaller gains are required to affect the poles.

In the rest of this chapter we obtain the eigenvalue gain sensitivities in terms of the flexible dynamics eigen–structure. Then the optimization problem is formulated as a continuous *minmax* problem. The design of a flexible–link planar arm is studied and the resulting optimized arm along with simulation results are presented.

6.2. Flexural Dynamics and Eigenvalue Sensitivities

Consider the dynamics of a flexible manipulator obtained by the Lagrangian method and given by [29]

$$M(q,\delta)\begin{bmatrix} \ddot{q} \\ \ddot{\delta} \end{bmatrix} + \begin{bmatrix} f_1(q,\dot{q}) + g_1(q,\dot{q},\delta,\dot{\delta}) \\ f_2(q,\dot{q}) + g_2(q,\dot{q},\delta,\dot{\delta}) + K\delta \end{bmatrix} = \begin{bmatrix} u \\ 0 \end{bmatrix} \quad (6.1)$$

where q is the $n \times 1$ vector of joint variables, δ is the $m \times 1$ vector of deflection variables, f_1, f_2, g_1, and g_2 are the terms due to gravity, Coriolis, and centripetal forces, M is the positive–definite mass matrix, K is the positive–definite stiffness matrix, and u is the $n \times 1$

vector of input torques (clamped mode shapes have been assumed).

Let us define $H(q, \delta) = M^{-1}(q, \delta) = \begin{bmatrix} H_{11} & H_{12} \\ H_{21} & H_{22} \end{bmatrix}$. Then the non-

linear dynamics of the flexible modes can be written as

$$
\begin{aligned}
\ddot{q} &= -H_{11}(f_1(q, \dot{q}) + g_1(q, \dot{q}, \delta, \dot{\delta})) \\
&- H_{12}(f_2(q, \dot{q}) + g_2(q, \dot{q}, \delta, \dot{\delta})) + K\delta) + H_{11}(q, \delta)u \qquad (6.2) \\
\ddot{\delta} &= -H_{21}(q, \delta)(f_1(q, \dot{q}) + g_1(q, \dot{q}, \delta, \dot{\delta})) \\
&- H_{22}(q, \delta)(f_2(q, \dot{q}) + g_2(q, \dot{q}, \delta, \dot{\delta})) + K\delta) + H_{21}(q, \delta)u \quad (6.3)
\end{aligned}
$$

Note that the dynamics (6.3) are generally *fast* as compared to the dynamics of the rigid–body modes governed by (6.2). Here, in order to capture the dominant properties of the above dynamics and to simplify our analysis, let us linearize the above system around a nominal configuration ($q = q_0$, $\dot{q} = 0$, $\delta = 0$, $\dot{\delta} = 0$) with a nominal input u_0. This will result in

$$
\begin{aligned}
\Delta\ddot{q} &= H_{11}(q_0, 0)\Delta u - H_{12}(q_0, 0)K\Delta\delta & (6.4) \\
\Delta\ddot{\delta} &= -H_{22}(q_0, 0)K\Delta\delta + H_{21}(q_0, 0)\Delta u. & (6.5)
\end{aligned}
$$

Now letting $x_f^T = [\delta^T \ \dot{\delta}^T]$ and $u_f = \Delta u$, we have from (6.5)

$$
\dot{x}_f = A_f x_f + B_f u_f \qquad (6.6)
$$

with A_f and B_f given by

$$
A_f = \begin{bmatrix} 0 & I \\ -H_{22}K & 0 \end{bmatrix}, \quad B_f = \begin{bmatrix} 0 \\ H_{21} \end{bmatrix} \qquad (6.7)
$$

where the dependencies on q_0 have been dropped for simplicity. Now consider the output as the vector of joint variables and let us define $y_f^T = [\Delta q^T \ \Delta \dot{q}^T]$. With this choice of output, the dynamics represented by (6.5) characterize the internal dynamics of the full-order system. The *zero–dynamics* can therefore be obtained by considering (6.5) when y_f is set to zero. Using the relationship $M_{22}^{-1} = H_{22} - H_{21}H_{11}^{-1}H_{12}$, the zero dynamics of (6.4), (6.5) take the form

$$
\dot{z} = A_z z \qquad (6.8)
$$

where $z^T = [\delta^T \ \dot{\delta}^T]$ and

$$
A_z = \begin{bmatrix} 0 & I \\ -M_{22}^{-1}K & 0 \end{bmatrix}. \qquad (6.9)
$$

Note that the inherent structural damping present in the actual arm will ensure stability of the internal dynamics. A desirable characteristic for a flexible–link system is to have high modal controllability at the system open-loop poles and reduced controllability at (or near) the system transmission zeros.

6.2.1.   Controllability and Sensitivity of Flexural Modes

In this section, we will obtain a controllability measure for the flexible system dynamics. Towards this end, consider system matrices A ($N\times N$), B ($N \times M$) and C ($L \times N$) corresponding to a MIMO system with N states, M inputs and L outputs. Let $A \in R^{N\times N}$ have N distinct eigenvalues and λ_i be the ith eigenvalue associated with the left and right eigenvectors f_i and e_i, respectively.

In [102], controllability and observability measures are defined for each mode λ_i based on the Frobenius norms of matrices $\Phi_B = adj(\lambda_i I - A)B$ and $\Phi_C = Cadj(\lambda_i I - A)$. Theoretically, the mode λ_i is uncontrollable (unobservable) if all the elements of Φ_B (Φ_C) are zero. The above norms measure how far the system is from uncontrollability or unobservability (for which these norms have zero magnitudes). It is shown in [102] that the controllability measure is given by

$$m_{ci} =\mid \pi_i \mid [f_i^* B B^T f_i]^{1/2} \tag{6.10}$$

and the observability measure is given by

$$m_{oi} =\mid \pi_i \mid [f_i^* f_i e_i^* C^T C e_i]^{1/2} \tag{6.11}$$

where

$$\pi_i = \prod_{j=1,\ j\neq i}^{N} (\lambda_i - \lambda_j).$$

Now let an output feedback control strategy with a feedback control gain matrix $G := \{G_{ij}\}$ ($i = 1,\cdots,\ M,\ j = 1,\cdots,\ L$) be used, resulting in the closed–loop system matrix $A+BGC$. To consider the effect on eigenvalues, the following sensitivity matrix can be obtained

$$S_i := \left[\begin{array}{ccc} \frac{\partial \lambda_i}{\partial G_{11}} & \cdots & \frac{\partial \lambda_i}{\partial G_{1L}} \\ \vdots & \ddots & \vdots \\ \frac{\partial \lambda_i}{\partial G_{M1}} & \cdots & \frac{\partial \lambda_i}{\partial G_{ML}} \end{array} \right] = \frac{B^T f_i (Ce_i)^*}{f_i^* e_i}. \tag{6.12}$$

The Frobenius norm of S_i, i.e., $\sigma_i = \|S_i\|_F$ can now be obtained as

$$\sigma_i = \sqrt{trace\{Ce_i f_i^* B)^*(Ce_i f_i^* B)\}}. \tag{6.13}$$

Now if e_i and f_i are scaled such that $f_i^* e_i$ and $e_i^* e_i$ are equal to one and $C = I$ (corresponding to state feedback of flexible modes) we have

$$\sigma_i = \sqrt{f_i^* BB^T f_i}. \tag{6.14}$$

Now let us consider the system given by (6.6) and perform the transformation

$$x_f = \begin{bmatrix} T & 0 \\ 0 & T \end{bmatrix} X \tag{6.15}$$

where T is a transformation matrix that diagonalizes $H_{22}K$. Note that since H_{22} and K are positive–definite, all the eigenvalues of $H_{22}K$ are positive (see e.g. [33]) and without loss of generality can be considered to be distinct. We then have

$$\dot{X} = AX + Bu \tag{6.16}$$

where

$$A = \begin{bmatrix} 0 & I \\ -\Lambda & 0 \end{bmatrix}, \quad B = \begin{bmatrix} 0 \\ T^{-1}H_{21} \end{bmatrix} \tag{6.17}$$

and $\Lambda = T^{-1}H_{22}KT = diag(\omega_{p1}^2, \cdots, \omega_{pm}^2)$.

For the above representation, the normalized left and right eigenvectors corresponding to $j\omega_{pi}$ ($i = 1, \cdots, m$) are

$$f_i^* = [\frac{\sqrt{1+\omega_{pi}^2}}{2}v_i \quad \frac{-j\sqrt{1+\omega_{pi}^2}}{2\omega_{pi}}v_i]$$

$$e_i^* = [\frac{1}{\sqrt{1+\omega_{pi}^2}}v_i \quad \frac{-j\omega_{pi}}{\sqrt{1+\omega_{pi}^2}}v_i] \tag{6.18}$$

where v_i is a $1 \times m$ vector whose ith element is 1 and all other elements are zero.

Now representing $T^{-1}H_{21}$ in (6.17) in the form

$$T^{-1}H_{21} = \begin{bmatrix} \gamma_{p1}^T \\ \gamma_{p2}^T \\ \vdots \\ \gamma_{pm}^T \end{bmatrix} \tag{6.19}$$

and using (6.10) we have

$$m_{ci} = (\prod_{j=1,\ j\neq i}^{m} |(\omega_j^2 - \omega_i^2)|)\sqrt{(1+\omega_i^2)\gamma_{pi}^T\gamma_{pi}}. \qquad (6.20)$$

Similarly σ_i for the i-th pole (denoted by σ_{pi}) is obtained from (6.14) as

$$\sigma_{pi} = \sqrt{\frac{1+\omega_{pi}^2}{4\omega_{pi}^2}\gamma_{pi}^T\gamma_{pi}}. \qquad (6.21)$$

Note that we are interested in having lower sensitivities when the closed–loop system poles are near the transmission zeros. Therefore σ_{pi} should be small at zero locations. A similar norm relationship can be found for the zeros by replacing ω_i, ω_j in (6.20) by the system transmission zeros, i.e.,

$$\sigma_{zi} = \sqrt{\frac{1+\omega_{zi}^2}{4\omega_{zi}^2}\gamma_{pi}^T\gamma_{pi}}. \qquad (6.22)$$

The above expressions will be used in Section 6.3 for formulating the *minmax* optimization problem.

One point should be mentioned in comparing these measures with previously obtained measures such as those in [101]. Considering the system given by (6.16)-(6.17) and γ_{pi} terms given by (6.19), it is noted that the magnitudes of pole frequencies can play significant roles in determining controllability properties. For example, note from (6.21) that if ω_i is small the sensitivity is critically dependent on ω_i. On the other hand, large values of ω_i do not have much effect on (6.21). However, for flexible structure systems, a major concern is the smallest pole and zero frequencies since they can adversely affect system performance. The frequency terms are missing in the measures obtained in [101] due to neglecting the effect of the A matrix in that study.

6.2.2. *Pole–zero Relationships*

In this section, we consider the flexural pole–zero frequency relationships since they are the main factors affecting the closed–loop system behavior in terms of the achievable bandwidth and speed of operation. The fast system poles and zeros are obtained from A_f and A_z

given by (6.7) and (6.8), respectively. More specifically, the poles are associated with the matrix $P_1 = H_{22}K$ and the zeros are associated with the matrix $P_2 = (H_{22} - H_{21}H_{11}^{-1}H_{12})K$. Using the fact that H and K are positive–definite matrices, it can be easily shown th at P_1 and P_2 have real positive eigenvalues and real eigenvectors (see e.g., [33]). The matrix P_2 can be considered as a perturbation of P_1 by $H_{21}H_{11}^{-1}H_{12}K$. In the following, we will show that the eigenvalues of P_2 are smaller than the corresponding eigenvalues of P_1.

Towards this end, first consider an arbitrary matrix $P \in R^{n \times n}$ with distinct eigenvalues and with its i–th eigenvalue and eigenvector given by λ_i and x_i, respectively. Now consider the eigen-structure of the perturbed matrix $P - \varepsilon Q$, where $Q \in R^{n \times n}$ and ε is a small positive value. Representing the eigenvector and eigenvalue of the perturbed system by $x_i(\varepsilon)$ and $\lambda_i(\varepsilon)$ we have [105]

$$x_i(\varepsilon) \;=\; x_i + \sum_{j=1,\, j \neq i}^{n} (-\varepsilon t_{j1} + \varepsilon^2 t_{j2} - O(\varepsilon^3)) x_j \qquad (6.23)$$

$$\lambda_i(\varepsilon) \;=\; \lambda_i - k_{i1}\varepsilon + k_{i2}\varepsilon^2 - O(\varepsilon^3) \qquad (6.24)$$

where t_{j1}, t_{j2}, and k_{i1}, k_{i2} are the first two constant terms in the Taylor series expansions of $x_i(\varepsilon)$ and $\lambda_i(\varepsilon)$, respectively. Furthermore, both the power series in ε are convergent. Starting from $(P - \varepsilon Q)x_i(\varepsilon) = \lambda_i(\varepsilon)x_i(\varepsilon)$, equating similar powers in ε, and using the fact that each left eigenvector y_i is orthogonal to the right eigenvector x_j for $i \neq j$, i.e., $y_i^T x_j = 0$, we have [105]

$$\frac{\partial \lambda_i}{\partial \varepsilon} = -k_{i1} = -\frac{y_i^T Q x_i}{y_i^T x_i}. \qquad (6.25)$$

Next, we will show that $\partial \lambda_i / \partial \varepsilon$ is negative if $P = H_{22}K$ and $Q = DK$, where H_{22} and K are the positive–definite mass and stiffness matrices as defined in (6.1) and (6.3), and $D = H_{21}H_{11}^{-1}H_{12}$ is a positive semi–definite matrix. Towards this end, now let x_i be the i–th right eigenvector of $H_{22}K$ (unperturbed system with $\varepsilon = 0$, cf. equation (6.23)), i.e., $H_{22}Kx_i = \lambda_i x_i$, where the λ_i's are all positive [33]. Therefore we can write

$$Kx_i = \lambda_i H_{22}^{-1} x_i. \qquad (6.26)$$

Assuming that the λ_i's are distinct, the left eigenvector y_i^T is given by $y_i^T = x_i^T H_{22}^{-1}/h_i$, where $h_i = x_i^T H_{22}^{-1} x_i = x_i^T K x_i / \lambda_i$. This can be easily evaluated by checking the relationship $y_i^T H_{22} K = \lambda_i y_i^T$ for $y_i^T = x_i^T H_{22}^{-1}/h_i$ and noting that the eigenvectors are unique up to a scalar multiple.

Now substituting for y_i^T in (6.25) and using $x_i^T H_{22}^{-1} = (Kx_i)^T/\lambda_i$ yields

$$\frac{\partial \lambda_i}{\partial \varepsilon} = -\frac{1}{h_i}(Kx_i)^T D(Kx_i) \le 0. \tag{6.27}$$

The equality condition in (6.27) is possible only if $Kx_i \in \mathcal{N}(H_{12})$. Now, considering (6.27), since λ_i is a decreasing function of ε it follows that $\lambda_i |_{\varepsilon=0} \ge \lambda_i |_{\varepsilon=1}$, or

$$\lambda_i(H_{22}K) \ge \lambda_i((H_{22} - H_{21}H_{11}^{-1}H_{12})K). \tag{6.28}$$

The equality sign corresponds to the case where a pole-zero cancellation occurs. However, when D is positive definite, then the inequality in (6.29) holds which implies that

$$\lambda_i(H_{22}K) > \lambda_i((H_{22} - H_{21}H_{11}^{-1}H_{12})K). \tag{6.29}$$

The implication of this result is that the first zero always occurs before the first pole when the output is taken as the joint variable. Increasing the first zero is important because it can increase the achievable closed-loop system bandwidth. The above issue has not been considered in the current literature [82], [21], [22], [71].

The above result is however compatible with some results that have been published in the literature concerning flexible space structures. In [106] the pole-zero locations are obtained for a flexible space structure and it is concluded that no zero has an imaginary part smaller than the smallest imaginary part of any pole. This result may seem to contradict what we have derived in this section. However, note that in the case of a flexible-link robot, one has also rigid-body degrees of freedom with poles located at $s = 0$. In our study, the pole–zero relationships are considered by decomposing the full dynamics into the slow dynamics, which correspond to $2n$ poles at $s = 0$, and the fast dynamics given by (6.6). Considering the dynamics of this system, the smallest poles are at the origin which occur before the first zero of the system. In this sense, our results are compatible with the results given in [106].

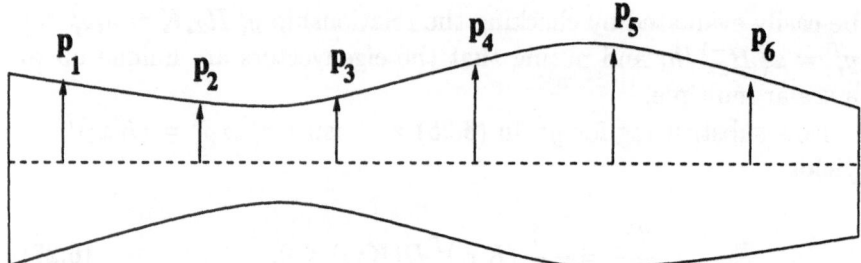

Figure 6.1. The parameter vector p determines the shape of the arm through B-splines.

6.3.   Multi–objective Optimization

As described in the previous section, several characteristics of the plant dynamics may be desirable when dealing with the geometric design of the arm. First, the smallest zero should be large enough so that higher bandwidths can be achieved. Also, we are interested in having higher eigenvalue sensitivities at the open–loop pole locations and reduced sensitivities at locations near the transmission zeros of the system. Towards this end, let us consider the following vector performance index

$$y(p) = \left[\begin{array}{ccccccc} \frac{1}{\omega_{z1}(p)} & \frac{1}{\sigma_{p1}(p)} & \cdots & \frac{1}{\sigma_{pr}(p)} & \sigma_{z1}(p) & \cdots & \sigma_{zs}(p) \end{array} \right]_{l \times 1}^{T} (6.30)$$

where $p^T = [p_1 \ p_2 \ \cdots \ p_r]$ is a parameter vector related to the geometric shapes of the arm as shown in Figure 6.1. For example p may consist of the parameters of the B-splines that determine the geometric shape of the arm. In Figure 6.1 changing the elements of the parameter vector corresponds to changing parameters in the B-splines which will change the shape of the arm.

Now suppose that p_0 is the initial parameter vector and $c = y(p_0)$. Let us normalize each $y_i(p)$ $(i = 1, \ \cdots, \ l)$ to its initial value c_i and define the following *minmax* problem

$$\underset{p}{minimize} \quad max(\frac{y_i(p)}{c_i}) \mid_{i=1,\cdots,l} .$$

$$(6.31)$$

The constraints for this problem can be the inertia experienced by the actuator or the mass of the link. At each time, the optimization algorithm deals with one objective at a time; the one with the largest ratio. However, the *minmax* function in (6.31) is discontinuous. To turn it into a smooth function we use the technique introduced in [103] which replaces the discrete *max* function by a continuous function of the form

$$\bar{\alpha}(p) = \frac{1}{\rho} ln\{\sum_{i=1}^{l} exp[\rho \frac{y_i(p)}{c_i}]\} \qquad (6.32)$$

for some $\rho > 0$. It is shown in [103] that $max(\frac{y_i(p)}{c_i}) \leq \bar{\alpha}(p) \leq max(\frac{y_i(p)}{c_i}) + \frac{ln(l)}{\rho}$. The choice of ρ therefore depends on the number of objectives l. For example if $l = 3$, then choosing $\rho = 20$ will result in an error of about 5 percent between $\bar{\alpha}(p)$ and $max(\frac{y_i(p)}{c_i})$, when $\bar{\alpha} \approx 1$. To improve the optimization process, the normalizing vector c can be updated to the last value of $y(p)$ from time to time, if $\bar{\alpha}(p)$ is to be reduced further [103].

6.4.   Optimization Procedure and Discussion of Results

In this section, we consider the design of a single–link flexible arm using the optimization scheme described in the previous section. For illustrative purposes, the arm is modeled using the assumed modes method with clamped–free mode shapes by employing the symbolic manipulation software *MAPLE* [35]. A more exact solution may, however, be carried out using finite–elements. The link is assumed to be made of aluminum with a $1kg$ payload at its end effector. A major consideration in the design of the mechanical structure should be directed towards reducing the torsional and lateral vibration effects since the conventional control inputs cannot directly affect these vibrations. Towards this goal, it can be shown (cf. e.g. [29]) that making the ratio of the cross–sectional height to its thickness large reduces the potential energy contribution of these deflections by a factor proportional to the inverse of the square of this ratio. A width to thickness ratio of 20 is used in the design.

In order to obtain smooth shapes for the link profile, the optimization problem can be formulated in terms of certain parameters defining *B–form* splines, or *B*–splines [83], [84]. A spline is defined by its (nondecreasing) *knot* sequence and its *B*–spline coefficient sequence. The elements of the knot sequence can be selected such that certain smoothness conditions are satisfied at each knot (determined by the multiplicity of each knot and the order of the B–splines). Thus, considering points within a certain interval (a certain knot sequence), the B–spline coefficients model the function they represent. For a specific knot sequence and by considering these coefficients as parameters, the geometric description of the link shapes are then at our disposal. Subsequently, the mass and stiffness matrices required in the optimization expressions are obtained. Thus the optimization problem may be formulated in terms of the B–spline coefficients that represent the parameter vector p in (6.31). Towards this goal the mass and stiffness matrices $M(q,0)$ and K which are functions of the parameter vector p are calculated. For example the first elements of the mass and stiffness matrices, denoted by M_{11} and K_{11}, respectively, are obtained from

$$M_{11}(p) = M_p L^2 + J_p + J_h + \int_0^L \rho A(x,p) x^2 dx$$

$$K_{11}(p) = E \int_0^L \phi(x)^2 I_o(x,p) dx$$

where M_p and J_p are the mass and inertia of the payload, J_h is the inertia of the hub, L is the length of the link, $A(x,p)$ is the cross sectional area of the link, I_o is the area moment of inertia about the neutral axis of the beam, E is Young's modulus of elasticity, $\phi(x)$ is the modal shape function of the first mode, and ρ is the mass density of the link. The following objective function was selected for the minmax optimization problem

$$Y(p) = \frac{1}{20} ln\{ exp(\frac{20\omega_{zmin,init}}{\omega_{zmin}(p)}) \quad + \quad exp(\frac{20\sigma_{p1,init}}{\sigma_{p1}(p)})$$
$$+ \quad exp(\frac{20\sigma_{z1}(p)}{\sigma_{z1,init}}) \} \quad (6.33)$$

where $\sigma_{z1,init}$ and $\sigma_{p1,init}$ are the initial sensitivities of the first zero and pole, respectively, and $\omega_{zmin,init}$ is the initial frequency of the first (minimum) zero. The constraint for the above objective function is that the total inertia experienced by the motor is to be less than or equal to that of the uniform case.

Figure 6.2 shows the thickness profiles before and after the optimization process. The lower and upper limits of the thickness profiles (the domain of optimization) are selected as one–third (0.2333 mm)

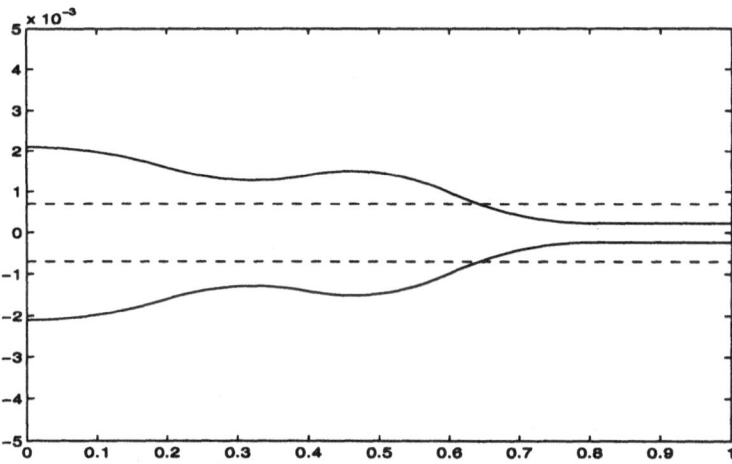

Figure 6.2. Thickness profile of the link (x–y axis in m) (optimized case: —, uniform and non–optimized case: – – –). The knot sequence is $\{0, 0, 0, l_2/5, 2l_2/5, ..., l_2, l_2, l_2\}$ and seven coefficients for the parameter vector are used.

and three times (2.1000 mm) the initial value of the thickness at the starting point (0.7000 mm), respectively. The lower limit may, however, be restricted by physical constraints such as the maximum allowable stress (cf. e.g. [86]). The vector optimization index is selected to improve the characteristics of the system corresponding to the first mode natural frequency, its gain sensitivity and the first zero gain sensitivity. The constrained optimization routine in the Matlab Optimization Toolbox is used to find the optimum shape. The algorithm is based on a sequential quadratic programming method.

The numerical values of parameters used in the optimization are as follows: $L = 1$ m, $M_p = 1$ kg, $J_h = 3 \times 10^{-5}$ kgm^2, $J_p = 0.002$ kgm^2, $\rho = 2700$ kg/m^3, $E = 6.93 \times 10^{10}$ N/m^2.

6.4.1. Performance Comparison between Non–uniform and Uniform Arms

In this section we compare the performance of the flexible–arm from different aspects before and after the optimization process. The first zero (with joint output) is moved from 1.13 to 5.1 rad/s. In terms of the achievable bandwidth, this means an improvement of nearly *five* times. The first pole sensitivity measure σ_{p1} is initially 330.3 which is changed to 1424, i.e., an increase by a factor of 4. The zero sensitivity measure σ_{z1} associated with the first zero was decreased from

6910 to 1104. The smallest natural frequency of the uniform arm is 17.54rad/s which is increased to 32.47 rad/s. The moment of inertia at the base of the flexible-link arm is the same for the optimized and non-optimized cases and is equal to $3.52 \times 10^{-2} kgm^2$. This is because the constraint is active after the optimization is terminated. The increased pole sensitivity means that it is possible to move the eigenvalues more easily towards the left half of the complex plane. On the other hand, the decreased zero sensitivity implies that poles are not easily attracted by zeros.

Next we study the performance of the optimized flexible-link system when a conventional control scheme for a *rigid* manipulator is applied to the system. To compare the dynamic behavior of the optimized and uniform cases, let us consider the joint position tracking control of a reference trajectory by using the *computed torque* control law based on the rigid part of the dynamic equations given by the following control law

$$u = I_0(\ddot{q}_r + K_d(\dot{q}_r - \dot{q}) + K_p(q_r - q)) \qquad (6.34)$$

where $I_0 = 1.037\ kgm^2$ is the rigid-body moment of inertia of the flexible beam about its hub, and q_r, $\dot{q}_r$, and $\ddot{q}_r$ represent the reference trajectory. Note that in the above control law we are only using the joint position and velocity information and no information regarding flexural variables is used. However, the complete nonlinear dynamic equations are taken into account using two flexible modes. The results are shown in Figures 6.3 and 6.4 for $K_p = K_d = 4$.

The sluggish response of the uniform arm is due to its smaller zero when compared to the optimized non–uniform arm. In terms of the torque requirements, it seems that the non-optimized link requires less torque initially. However, this behavior is a result of the large errors that are introduced due to the control law (6.34) at the beginning of the trajectory. A physical justification for this behavior is as follows: At the beginning of the trajectory, this control torque causes larger and more rapid joint motion for the flexible (uniform) link than for its flexible (non-uniform) counterpart. This results in q leading q_r as observed from the simulation results in Figures 6.3 and 6.4. Thus the terms $\dot{q}_r - \dot{q}$ and $q_r - q$ are negative at the beginning of the trajectory, which tends to initially reduce the torque input u given by (6.34). This reduction is greater for the flexible (uniform) link because of its larger and more rapid initial joint motion. Because of the link flexibility, the other parts of the link experience

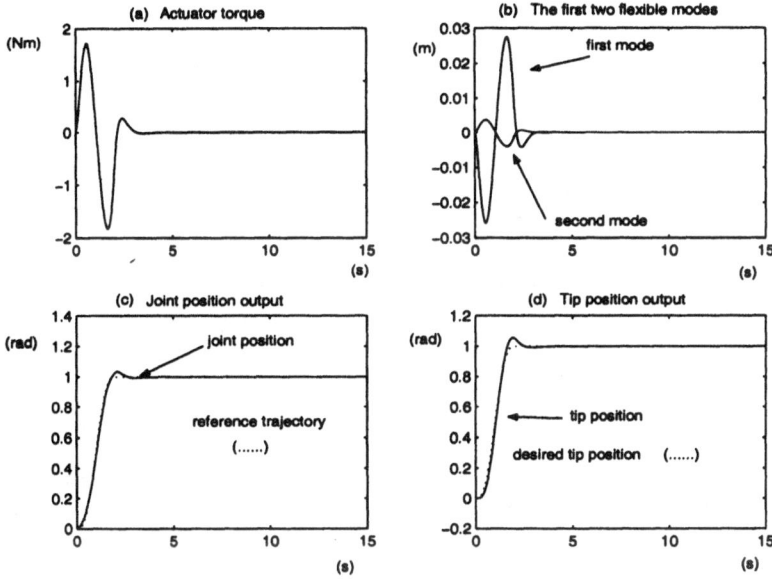

Figure 6.3. Dynamic behavior for the optimized non–uniform manipulator under joint feedback control for $K_p = K_d = 4$.

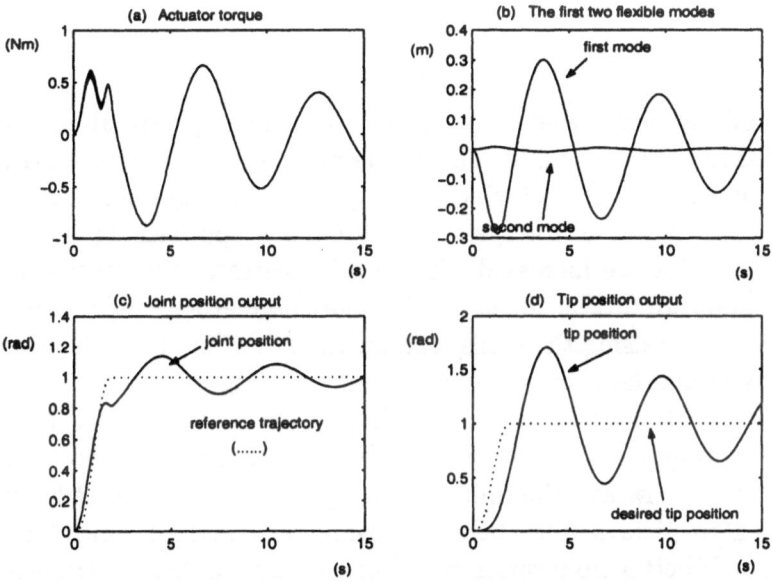

Figure 6.4. Dynamic behavior for the uniform manipulator under joint feedback control for $K_p = K_d = 4$.

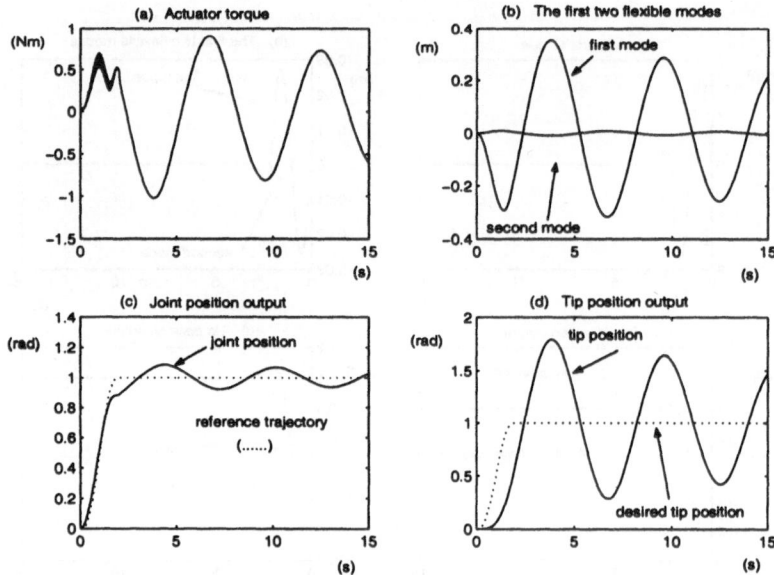

Figure 6.5. Dynamic behavior for the non-optimized uniform manipulator under joint feedback control for $K_p = 9$, $K_d = 6$.

some delay before they move. During this delay potential energy is stored in the flexible link. The oscillatory response in Figure 6.4 has a frequency of 1.13 rad which corresponds to the system zero. This is a consequence of the system poles quickly approaching the zeros as PD gains are increased. Figure 6.5 illustrates the results when PD gains are further increased, i.e. more oscillatory tip-position behavior. Decreasing the gains will, on the other hand, result in poor behavior and large errors as is evident from Figure 6.4d.

Clearly, the optimized manipulator shows much better performance when a conventional joint feedback control strategy is employed. In contrast, the uniform non-optimized arm cannot be controlled to yield similar behavior using conventional control techniques. If better tip position tracking is desired, either the trajectory should be slowed down, which results in performance degradation, or more advanced control techniques have to be employed (cf., [70], [69]).

6.5.   Conclusion

The results for different simulations of an optimized non-uniform link show that improvements are achieved over the case of a manipulator with a uniform link. The improvements are obtained on the smallest transmission zero, the lowest natural frequency of flexible modes, modal sensitivities at open–loop pole locations, and pole sensitivities at the locations of the transmission zeros. These improvements are achieved without exceeding the constraints on the inertia experienced by the motor. Because of the above requirements, the optimization parameters change in such a way that the mass matrix of the manipulator is not significantly different from the uniform case. However, the changes are mostly present in the stiffness matrix that is no longer diagonal for the non–uniform optimized manipulators (assuming that orthogonal shape functions are used). Finally, a comparison of the performance between the uniform manipulator and the optimized non–uniform manipulator shows that subject to the same control strategy, the optimized non–uniform manipulator exhibits improved dynamic behavior in a state or output feedback strategy. There is certainly a limit on the amount of improvement that may be achieved by the optimized structure. Further improvements can be accomplished by employing more advanced controllers as in [69] and [70].

7.

Concluding Remarks

It is now nearly more than a decade that researchers in control and robotics have been investigating the problem of controlling robotic mechanisms with considerable structural flexibilities. Nevertheless, it seems that, from a control and design perspective, this issue will remain an open research area for several years to come. As was mentioned earlier, several factors contribute to the complexity of the problem. The dynamic model complexity, uncertainties in modeling, ill–conditioned nonlinear dynamics, and nonminimum–phase characteristics are among the major factors.

In this monograph we have focused on two nonlinear model–based design methods for trajectory tracking control, and have considered a structural shape design. The control strategies were implemented and tested on an experimental two–link flexible setup that was constructed in the laboratory. This system possesses interesting nonlinear and nonminimum–phase features that can be found in other systems such as aircraft, underwater vehicles, large flexible satellites, and other under–actuated mechanical systems. Although the control goals may be different, many similarities exist between these systems. Therefore the control methods can be applied to other cases once appropriate modifications are made. The main advantages of the control strategies studied in this monograph over conventional control methods are robust closed–loop performance and sufficiently small tip–position tracking errors. However, the price to be paid is more control complexity and the need for faster processors for real–time computation. Based on the experience gained in the course of this research, the following routes may be taken for further work.

7.1.   Control Using Integral Manifolds

One problem with this control strategy is the complexity of the controller. This is a major problem, specifically if the number of links is increased. One way to deal with this issue is to use some kind of function approximation by employing neural networks or spline functions. Splines can be used to approximate functions that cannot be represented in closed-form and can be used to represent nonlinearities in plant models and nonlinear controllers [89]. Further improvement of this control strategy lies in devising or modifying the control law to deal with parametric uncertainties and parameter variations. In this regard, adaptive techniques may be considered. However, these techniques may further raise complexity issues in control implementation and should be studied along with the function approximation mentioned earlier.

7.2.   Input–output Decoupling Control

The control strategy based on decoupling and using observers for estimating flexural rates is generally less complicated, in terms of implementation, compared to the integral manifold method. One problem that arises here is that the ill–conditioning in the mass matrix and parametric uncertainties can cause instability [91]. It may be possible to resolve these issues using dynamic decoupling [68], [72], and robust designs such as sliding techniques [75].

7.3.   Structure Design

As discussed previously, a major goal in structural shape design is to achieve better robustness in a closed–loop control system. It is possible that a better optimization index can be defined if controllability and observability properties can also be incorporated in shape design. Increasing the structural natural frequencies has to be kept as one of the optimization goals. In addition, as the number of optimization goals increases, consideration should be given to employing appropriate schemes for multi–objective optimization for the multi-link manipulator.

In this monograph, we have only considered mechanical shape design. While this is an important aspect, one should not overlook *smart* techniques that help to reduce the degree of difficulty of the control problem through the choice/design of appropriate material type, actuators and sensors (see e.g., [28]). A successful solution should be an integration of control and structure/material design. As far as the control design is concerned, one may follow classical techniques, intelligent methods (fuzzy logic, neural networks, neuro–fuzzy), or a combination of the two. A successful control strategy should consider robustness to parametric uncertainties as well as higher frequency flexural modes while achieving desired performance.

Undoubtedly, more extensive benchmark comparisons have to be made before any conclusions can be drawn concerning the superiority of one approach over another.

Appendix A.
Stability Proofs

The stability analysis for the various key results obtained in this research are given in this appendix.

A.1. Proof of Theorem 2.1

The stability analysis follows along the lines developed in [53]. Consider the open–loop system (2.5)–(2.6) with the control laws obtained from (2.10), (2.11), (2.25), (2.35), (2.41), and (2.42), that is

$$
\begin{aligned}
u &= u_s(x, \varepsilon, t) + u_f(x, z) \\
u_s &= u_0 + \varepsilon u_1 + \varepsilon^2 u_2 \\
u_0 &= M_{110} v_0 + f_1(x_1, x_2) \\
u_1 &= M_{110} v_1 \\
u_2 &= M_{110}(v_2 + d(x_1, x_2, \dot{x}_2 \mid_{\varepsilon=0}, u_0, \dot{u}_0) - \Psi \ddot{h}_{10}) \\
\dot{v} &= A_v v + b_v(\ddot{y}_r, e, \varepsilon) \\
\varepsilon \dot{w} &= F(x_1) w + G(x_1) y_f \\
u_f &= M(x_1) w.
\end{aligned}
\tag{A.1}
$$

Substituting the above control laws into system (2.5)–(2.6) and adopting the same definitions for e and $\tilde{z}$ as before, after some algebraic manipulations, yields

$$\begin{aligned}
\dot{e} &= A_e e + b_e(x, \tilde{z}, w, v, \varepsilon, t) \\
\dot{v} &= A_v v + b_v(\ddot{y}_r, e, \varepsilon) \\
\varepsilon \dot{\eta} &= A_\eta(x_1)\eta + b_\eta(x, \tilde{z}, w, v, t)
\end{aligned} \tag{A.2}$$

where $\eta^T = [\tilde{z}^T \ w^T]$, and $b_v(\ddot{y}_r, e, \varepsilon)$ is the same as before, and

$$b_e(x, \tilde{z}, w, v, \varepsilon, t) =$$

$$\begin{bmatrix} 0 \\ -\frac{H_{120}K}{\lambda_{min}}\tilde{z}_1 + H_{110}M(x_1)w - \varepsilon(H_{110}G_{11} + H_{120}G_{21})\tilde{z}_2 + O_1(\varepsilon^2) \end{bmatrix}$$

$$b_\eta(x, \tilde{z}, w, v, t) = \begin{bmatrix} -\varepsilon^3 \dot{h}_{12} \\ -\varepsilon(H_{210}G_{11} + H_{220}G_{21})\tilde{z}_2 + O_2(\varepsilon^2) \\ \varepsilon^3 G(x_1)\Psi(h_{11} + \varepsilon h_{12}) \end{bmatrix}. \tag{A.3}$$

Noting that the matrices A_e, A_v, and $A_\eta(x_1)$ are all Hurwitz, we can write the following Lyapunov equations

$$\begin{aligned}
A_e^T P_e + P_e A_e &= -Q_e \\
A_v^T P_v + P_v A_v &= -Q_v \\
A_\eta^T(x_1)P_\eta(x_1) + P_\eta(x_1)A_\eta(x_1) &= -Q_\eta
\end{aligned} \tag{A.4}$$

where Q_e, Q_v, and Q_η are positive definite symmetric $O(1)$ matrices. Choosing the positive definite Lyapunov function candidate

$$V = e^T P_e e + \hat{v}^T P_v \hat{v} + \varepsilon \eta^T P_\eta(x_1)\eta \tag{A.5}$$

where $\hat{v} = \varepsilon^2 v$ [43], and computing the derivative of V along the trajectories of (A.2) yields

$$\begin{aligned}
\dot{V} &= -e^T Q_e e - \hat{v}^T Q_v \hat{v} - \eta^T Q_\eta \eta + 2e^T P_e b_e \\
&+ 2\varepsilon^2 \hat{v} P_v b_v + 2\eta^T P_\eta(x_1)b_\eta + \varepsilon\eta^T \dot{P}_\eta(x_1)\eta.
\end{aligned} \tag{A.6}$$

A detailed inspection of the terms b_e, b_v, and b_η reveals that on bounded regions $\Omega_1, \Omega_2, \Omega_3 \subset \mathbf{R}^{4n+2m+l}$ around the origin of the (e, v, η) state space, and by assuming a C^2 desired reference trajectory, we may write

$$\begin{aligned}
\|b_e\| &\leq \gamma_E\|\eta\| + \varepsilon^2(l_1 + l_{11}\|e\| + l_{12}\|\eta\|), &\forall (e, v, \eta) \in \Omega_1 \\
\|b_v\| &\leq \gamma_r + \gamma_{pd}\|e\|, &\forall (e, v, \eta) \in \Omega_2 \\
\|b_\eta\| &\leq \varepsilon(\gamma_\eta + \gamma_{\eta 1}\|e\|)\|\eta\| + \varepsilon^2(l_2 + l_{21}\|e\| + l_{22}\|\eta\|), \\
&\forall (e, v, \eta) \in \Omega_3
\end{aligned} \tag{A.7}$$

where l_1, l_2, l_{11}, l_{12}, l_{21}, l_{22} are upper bound constants specified in the regions Ω_1 and Ω_3. Furthermore

$$
\begin{aligned}
\gamma_E &= \|[-\frac{H_{120}K}{\lambda_{min}}, \quad -\varepsilon(H_{110}G_{11} + H_{120}G_{21}), \\
&\qquad H_{110}M(x_1)]\|_{max} \text{ on } \Omega_1 \\
\gamma_r &= \|\frac{A_1 A_2^{-1}}{\varepsilon^2}\ddot{y}_r\|_{max} \\
\gamma_{pd} &= \|\frac{A_1 A_2^{-1}}{\varepsilon^2}[K_p, \quad K_d]\| \\
\gamma_\eta &= \|H_{210}G_{11} + H_{220}G_{21}\|_{max} \text{ on } \Omega_3,\, e=0.
\end{aligned}
\tag{A.8}
$$

Similarly, for all $(e, v, \eta) \in \Omega_4 \subset \mathbf{R}^{4n+2m+l}$ we have

$$
\|\dot{P}_\eta(x_1)\| \le l_3
\tag{A.9}
$$

Thus defining

$$
\lambda_e := \lambda_{min}\{Q_e\}, \quad \lambda_v := \lambda_{min}\{Q_v\}, \quad \lambda_\eta := \lambda_{min}\{Q_\eta\}. \tag{A.10}
$$

and making use of (A.7) and (A.9) in (A.6) results in

$$
\dot{V} \le -[\|e\| \ \|\hat{v}\| \ \|\eta\|]\Lambda \begin{bmatrix} \|e\| \\ \|\hat{v}\| \\ \|\eta\| \end{bmatrix} + 2\alpha^T \begin{bmatrix} \|e\| \\ \|\hat{v}\| \\ \|\eta\| \end{bmatrix}, \quad \forall(e, v, \eta) \in \Omega_I
\tag{A.11}
$$

where $\Omega_I = \Omega_1 \cap \Omega_2 \cap \Omega_3 \cap \Omega_4$ and

$$
\begin{aligned}
\alpha^T &= \varepsilon^2[l_1\|P_e\| \ \gamma_r\|P_v\| \ l_2\|P_\eta\|] \\
\Lambda &= \begin{bmatrix} \Lambda_{11} & \Lambda_{12} & \Lambda_{13} \\ \Lambda_{12} & \Lambda_{22} & \Lambda_{23} \\ \Lambda_{13} & \Lambda_{23} & \Lambda_{33} \end{bmatrix}
\end{aligned}
$$

with Λ_{ij}, $i,\, j = 1, \cdots, 3$, given by

$$
\begin{aligned}
\Lambda_{11} &= \lambda_e - 2\varepsilon^2 l_{11}\|P_e\|, \quad \Lambda_{12} = -\varepsilon^2\gamma_{pd}\|P_v\|, \\
\Lambda_{13} &= -\gamma_E\|P_e\| - \varepsilon\gamma_{\eta_1}\|P_\eta\| - \varepsilon^2(l_{12}\|P_e\| + l_{21}\|P_\eta\|), \\
\Lambda_{22} &= \lambda_v, \quad \Lambda_{23} = 0, \Lambda_{33} = \lambda_\eta - \varepsilon(2\lambda_\eta\|P_\eta\| + l_3) - 2\varepsilon^2 l_{22}\|P_\eta\|.
\end{aligned}
$$

Noting that $\gamma_{pd}\|P_v\|$ will remain $O(1)$ as ε tends to zero, we conclude that if $\lambda_e\lambda_\eta > \gamma_E\|P_e\|^2$, then Λ will be positive–definite for all $\varepsilon \in$

$(0, \varepsilon_{max})$, where ε_{max} is the upper bound for ε. Assuming that ε lies in this interval, let us apply the coordinate transformation

$$\mathcal{X} = \begin{bmatrix} \|e\| \\ \|\hat{v}\| \\ \|\eta\| \end{bmatrix} - \Lambda^{-1}\alpha \qquad (A.12)$$

that when substituted in (A.11) yields

$$\dot{V} \leq -\mathcal{X}^T\Lambda\mathcal{X} + \alpha^T\Lambda^{-1}\alpha \qquad \forall(e, v, \eta) \in \Omega_I. \qquad (A.13)$$

Consider the case when $\mathcal{X}^T\Lambda\mathcal{X} = \alpha^T\Lambda^{-1}\alpha$. This is the equation of an ellipsoid in $\mathcal{X}$ coordinates. Furthermore, using the spectral theorem of linear algebra, Λ may be written as $\Lambda = Q^T\Lambda_d Q$, where Q is a matrix whose columns are the orthogonal eigenvectors of Λ, and Λ_d is a diagonal matrix of the eigenvalues of Λ. The equation of the ellipsoid may then be written as $(Q^T\mathcal{X})^T\Lambda_d(Q^T\mathcal{X}) = \alpha^T\Lambda^{-1}\alpha$. Thus the largest and smallest diagonals of the ellipsoid are given respectively by $\sqrt{\alpha^T\Lambda^{-1}\alpha/\lambda_{min}(\Lambda)}$ and $\sqrt{\alpha^T\Lambda^{-1}\alpha/\lambda_{max}(\Lambda)}$.

To show the boundedness of solutions, let us consider the following sets

$$\begin{aligned} \mathcal{R} &= \{(e, \hat{v}, \eta) \mid \mathcal{X}^T\Lambda\mathcal{X} \leq \alpha^T\Lambda^{-1}\alpha\} \cap \Omega_I \\ \mathcal{S} &= \{(e, \hat{v}, \eta) \mid V(e, \hat{v}, \eta) \leq c_1\} \subset \Omega_I \\ \mathcal{T} &= \{(e, \hat{v}, \eta) \mid V(e, \hat{v}, \eta) = c_2\} \subset \Omega_I \end{aligned} \qquad (A.14)$$

where $0 < c_2 \leq c_1$ and c_2 is the smallest constant such that $\mathcal{R} \subset \mathcal{T}$. Since the trajectory defined by y_r, $\dot{y}_r$, $\ddot{y}_r$ is bounded and Λ is positive–definite, $\mathcal{R}$ is uniformly bounded. If the initial state is outside $\mathcal{S} - \mathcal{R}$, where $\dot{V} \leq 0$, it follows that there exists a finite time t_f such that any solution starting from $\mathcal{S} - \mathcal{R}$, at $t > 0$, will enter $\mathcal{T}$ at t_f, and reside in $\mathcal{T}$ thereafter for all $t \geq t_f$. The residual set $\mathcal{T}$ encompasses the ellipsoid. Thus the size of the residual set is on the order of the maximum diameter of the ellipsoid. Since Λ is $O(1)$ and α is $O(\varepsilon^2)$, the residual set is of $O(\varepsilon^2)$. Therefore e, $\hat{v}$, and η remain bounded up to $O(\varepsilon^3)$ after t_f. Thus, v will be $O(1)$ in the steady state. This completes the proof of the theorem stated in Section 3.

A.2. Proof of Theorem 3.1

Consider the dynamic equations of the closed–loop system given by (3.22) and (3.24). Since A_E and $A_\Delta(q)$ are Hurwitz matrices we have the Lyapunov equations

$$A_E^T P_E + P_E A_E = -Q_E$$
$$A_\Delta^T(q) P_\Delta(q) + P_\Delta(q) A_\Delta(q) = -Q_\Delta \qquad (A.15)$$

where P_E, Q_E, $P_\Delta(q)$, Q_Δ are positive–definite matrices. Let us choose the positive–definite Lyapunov function candidate

$$V = E^T P_E E + \hat{\Delta}^T P_\Delta(q) \hat{\Delta} \qquad (A.16)$$

where $\hat{\Delta} = \epsilon_1 \Delta$ with ϵ_1 being a positive constant (typically less than one). Then $\dot{V}$ is given by

$$\dot{V} = -E^T Q_E E - \hat{\Delta}^T Q_\Delta \hat{\Delta} + 2 d_E^T P_E E + 2\epsilon_1 \hat{\Delta} P_\Delta G_\Delta + \hat{\Delta} \dot{P}_\Delta \hat{\Delta}. \quad (A.17)$$

Consider a continuous bounded reference trajectory (at least C^2) and a bounded region Ω_1 containing the origin of E and Δ. Then for all $(E, \Delta) \in \Omega_1 \subset \mathbf{R}^{m+n}$ we can write

$$\|a\| \le l_1 + l_2\|E\| + l_3\|\Delta\|, \quad \|I - B\hat{B}^{-1}\| \le \epsilon_2, \quad \|B\hat{B}^{-1}\|\|\Delta a\| \le \epsilon_3 \qquad (A.18)$$

where l_i's and ϵ_i's are certain bounds with ϵ_i's being typically small quantities. Further, let

$$\|[K_p \ K_d]\| = l_4, \quad \|[BK_\delta \ BK_{\dot{\delta}}]\| \le l_5, \quad \|\ddot{y}_r\| \le l_6 \qquad (A.19)$$

to get

$$\|d_E\| \le \epsilon_3 + \epsilon_2(l_1 + l_6) + \epsilon_2(l_2 + l_4)\|E\| + (l_5 + \epsilon_2 l_3)\frac{\|\hat{\Delta}\|}{\epsilon_1}. \quad (A.20)$$

Similarly, for all $(E, \Delta) \in \Omega_2 \subset \mathbf{R}^{m+n}$, where Ω_2 is a finite region containing the origin of (E, δ), we have

$$\|H_{210}\hat{B}_0^{-1}\| \approx 1, \quad \|H_{210}(-\hat{B}_0^{-1}\hat{a}_0 + B_0^{-1}a_0\| \le \epsilon_4(l_6 + l_7\|E\|),$$
$$\|O(\delta^2, q, \dot{q})\| \le l_8. \qquad (A.21)$$

Thus

$$\|G_\Delta\| \le l_6 + l_4\|E\| + \epsilon_4(l_6 + l_7\|E\|) + l_8, \quad \forall (E, \Delta) \in \Omega_2. \quad (A.22)$$

A detailed analysis also reveals that for all $(E, \Delta) \in \Omega_3 \subset \mathbf{R}^{m+n}$, where Ω_3 is a finite region containing the origin of (E, δ), we have

$\|\dot{P}_\delta(q)\| \le l_9$. Hence, letting $\Omega_i = \Omega_1 \cap \Omega_2 \cap \Omega_3$ and substituting the previous inequalities in $\dot{V}$ given by (A.17) and rearranging the right–hand side of the resulting inequality in terms of $\|E\|$ and $\|\Delta\|$ will, after some algebraic manipulations, yield

$$\dot{V} \le -[\|E\| \ \|\hat{\Delta}\|]\Lambda \left[\begin{array}{c} \|E\| \\ \|\hat{\Delta}\| \end{array} \right] + 2\gamma^T \left[\begin{array}{c} \|E\| \\ \|\hat{\Delta}\| \end{array} \right] \qquad \forall (E,\Delta) \in \Omega_i \quad (A.23)$$

where

$$
\begin{aligned}
\Lambda &= \left[\begin{array}{cc} \Lambda_{11} & \Lambda_{12} \\ \Lambda_{12} & \Lambda_{22} \end{array} \right], \\
\Lambda_{11} &= \lambda_e - 2\epsilon_2(l_2 + l_4)\|P_E\|, \quad \Lambda_{22} = \lambda_\Delta - l_9 \\
\Lambda_{12} &= \frac{(l_5 + \epsilon_2 l_3)\|P_E\|}{\epsilon_1} + \epsilon_1 \|P_\Delta\|(l_4 + \epsilon_4 l_7), \\
\gamma^T &= [(\epsilon_3 + \epsilon_2(l_1 + l_6))\|P_E\| \quad \epsilon_1\|P_\Delta\|(l_6 + l_8 + \epsilon_4 l_6)], \\
\lambda_e &= \lambda_{min}(Q_E), \quad \lambda_\Delta = \lambda_{min}(Q_\Delta).
\end{aligned}
$$

$$(A.24)$$

Now consider the case where there are no parameter errors. Provided that $\lambda_\Delta > l_9$, and $\lambda_e(\lambda_\Delta - l_9) > l_5\|P_E\|/\epsilon_1 + \epsilon_1 l_4\|P_\Delta\|$, the matrix Λ is positive definite. It then follows that for a certain range of parameter variations in this neighborhood, Λ remains positive definite. Assuming that this condition holds, let us define the sets

$$
\begin{aligned}
\mathcal{R} &= \{(E,\hat{\Delta}) \mid [\|E\| \ \|\hat{\Delta}\|]\Lambda[\|E\| \ \|\hat{\Delta}\|]^T \le 2\gamma^T[\|E\| \ \|\hat{\Delta}\|]^T\} \cap \Omega_i \\
\mathcal{S} &= \{(E,\hat{\Delta}) \mid V(E,\hat{\Delta}) \le c_1\} \subset \Omega_i \\
\mathcal{T} &= \{(E,\hat{\Delta}) \mid V(E,\hat{\Delta}) = c_2\} \subset \Omega_i
\end{aligned}
$$

$$(A.25)$$

where $0 < c_2 \le c_1$ and c_2 is the smallest constant such that $\mathcal{R} \subset \mathcal{T}$. Since the trajectory defined by y_r, $\dot{y}_r$, $\ddot{y}_r$ is bounded and Λ is positive–definite, $\mathcal{R}$ is uniformly bounded. If the initial state is outside $\mathcal{S} - \mathcal{R}$, where $\dot{V} \le 0$, it follows that there exists a finite time t_f such that any solution starting from $\mathcal{S} - \mathcal{R}$, at $t > 0$, will enter $\mathcal{T}$ at t_f, and remain in $\mathcal{T}$ thereafter for all $t \ge t_f$. The residual set $\mathcal{T}$ encompasses an ellipse in $[\|E\| \ \|\hat{\Delta}\|]$ coordinates with its diameters being $\sqrt{\gamma^T \Lambda^{-1} \gamma / \lambda_{min}(\Lambda)}$ and $\sqrt{\gamma^T \Lambda^{-1} \gamma / \lambda_{max}(\Lambda)}$. Since $\hat{\Delta} = \epsilon_1 \Delta$, it follows that Δ is of the order of $\frac{1}{\epsilon_1}$ on this residual set. It is possible to obtain an optimum value of ϵ_1 by setting the parameter errors to zero in Λ, which gives $\epsilon_1 = \sqrt{l_5 \|P_E\| / (l_4 \|P_\Delta\|)}$. Some qualitative robustness measures may be obtained from this analysis. In particular, making the norms of P_E, $K_\delta(q)$, $K_{\hat{\delta}}(q)$ small and reducing the parameter errors yield a better margin of robustness. Note however that the lower limit of $K_\delta(q)$, $K_{\hat{\delta}}(q)$ is not zero since A_Δ should be a Hurwitz matrix.

A.3. Proof of Theorem 4.1

In what follows a closed–loop stability analysis is established for the case of a full–order observer. A similar analysis can be given for the case where a reduced–order observer is used. Consider the dynamic equations of the closed–loop system given by (4.10)–(4.11), (4.21)–(4.22) and (4.23)–(4.24) which are repeated here for convenience

$$\dot{\tilde{\delta}} = A_{\tilde{\delta}}\tilde{\delta} + b_{\tilde{\delta}}(x, \hat{\delta}_2) \tag{A.26}$$

$$\dot{E} = A_E E + d_E(\alpha, x_c, t) \tag{A.27}$$

$$\dot{\Delta} = A_\Delta(q)\Delta + \begin{bmatrix} 0 \\ G_\Delta(x, \hat{\delta}, t) \end{bmatrix}. \tag{A.28}$$

Since $A_{\tilde{\delta}}$, A_E and $A_\Delta(q)$ are Hurwitz matrices, the following Lyapunov equations are satisfied

$$
\begin{aligned}
A_{\tilde{\delta}}^T P_{\tilde{\delta}} + P_{\tilde{\delta}} A_{\tilde{\delta}} &= -Q_{\tilde{\delta}} \\
A_E^T P_E + P_E A_E &= -Q_E \\
A_\Delta^T(q) P_\Delta(q) + P_\Delta(q) A_\Delta(q) &= -Q_\Delta
\end{aligned} \tag{A.29}
$$

where $P_{\tilde{\delta}}$, $Q_{\tilde{\delta}}$, P_E, Q_E, $P_\Delta(q)$, Q_Δ are symmetric positive–definite matrices. Let us choose the positive–definite Lyapunov function candidate

$$V = E^T P_E E + \hat{\Delta}^T P_\Delta(q)\hat{\Delta} + \tilde{\delta}^T P_{\tilde{\delta}}\tilde{\delta} \tag{A.30}$$

where $\hat{\Delta} = \epsilon \Delta$ with ϵ being a small positive scaling constant (typically less than one). The scaling factor is introduced as a result of from the stability requirement that E and $\tilde{\delta}$ converge to small values near zero while Δ remains bounded. Thus, the scaling factor allows us to show that E, $\tilde{\delta}$ and $\hat{\Delta}$ converge to small values, hence Δ converges to $\hat{\Delta}/\epsilon$.

Taking the time derivative of V yields

$$
\begin{aligned}
\dot{V} =\ & -E^T Q_E E - \hat{\Delta}^T Q_\Delta \hat{\Delta} - \tilde{\delta}^T Q_{\tilde{\delta}} + 2 d_E^T P_E E + 2 b_{\tilde{\delta}}^T P_{\tilde{\delta}} \tilde{\delta} \\
& + 2\epsilon \hat{\Delta} P_\Delta G_\Delta + \hat{\Delta} \dot{P}_\Delta \hat{\Delta}.
\end{aligned}
\tag{A.31}
$$

Consider a continuous bounded reference trajectory (at least C^2) and a bounded region Ω_1 containing the origin of $(E, \Delta, \tilde{\delta})$. Then for all $(E, \Delta, \tilde{\delta}) \in \Omega_1 \subset \mathbf{R}^{m+4n}$, from (4.22) we have

$$
\|d_E\| \leq \frac{\epsilon_0}{\epsilon} \|\hat{\Delta}\| + \epsilon_1 \|\tilde{\delta}\|
\tag{A.32}
$$

where ϵ_0 depends on $K_\delta(q)$ and $K_{\hat{\delta}}(q)$ and should be sufficiently small by proper choice of these matrices as we shall see shortly. The ϵ_1 term is affected by $\|K_d\|$ and the higher order terms in (4.22). Similarly, for all $(E, \Delta, \tilde{\delta}) \in \Omega_2 \subset \mathbf{R}^{m+4n}$, where Ω_2 is a finite region containing the origin of $(E, \Delta, \tilde{\delta})$, from (4.24) we have

$$
\|G_\Delta\| \leq l_{\Delta_1} + l_E \|E\| + \frac{\epsilon_2}{\epsilon} \|\hat{\Delta}\| + \epsilon_3 \|\tilde{\delta}\|, \quad \forall (E, \Delta, \tilde{\delta}) \in \Omega_2
\tag{A.33}
$$

where l_{Δ_1} is mainly affected by the reference trajectory and l_E is affected by K_p and K_d. Constants ϵ_2 and ϵ_3 correspond to $O(\delta)$ and $O(\delta \hat{\delta})$ terms in (4.24), respectively. A detailed analysis also reveals that for all $(E, \Delta, \tilde{\delta}) \in \Omega_3 \subset \mathbf{R}^{m+4n}$, where Ω_3 is a finite region containing the origin of $(E, \Delta, \tilde{\delta})$, we have $\|\dot{P}_\Delta(q)\| \leq l_{\Delta_2}$. This can be shown by noting that $P_\Delta(q)$ has a finite growth rate with respect to q (see e.g. [67], Problem 5.21) and the reference trajectory is bounded. Hence, letting $\Omega_i = \Omega_1 \cap \Omega_2 \cap \Omega_3$ and substituting the previous inequalities in $\dot{V}$ and rearranging the right–hand side of the resulting inequality in terms of $\|E\|$, $\|\Delta\|$ and $\|\tilde{\delta}\|$ after some algebraic manipulations, yields

$$
\dot{V} \leq -[\|E\| \ \|\hat{\Delta}\| \ \|\tilde{\delta}\|] \Lambda \begin{bmatrix} \|E\| \\ \|\hat{\Delta}\| \\ \|\tilde{\delta}\| \end{bmatrix} + 2\gamma^T \begin{bmatrix} \|E\| \\ \|\hat{\Delta}\| \\ \|\tilde{\delta}\| \end{bmatrix} \qquad \forall (E, \Delta, \tilde{\delta}) \in \Omega_i
\tag{A.34}
$$

where

$$
\begin{aligned}
\Lambda &= \begin{bmatrix} \lambda_e & \frac{\epsilon_0}{\epsilon} + l_E \epsilon \|P_\Delta\| & \epsilon_1 \|P_E\| \\ \frac{\epsilon_0}{\epsilon} + l_E \epsilon \|P_\Delta\| & \lambda_\Delta - l_{\Delta_2} - 2\|P_\Delta\|\epsilon_2 & \epsilon\epsilon_3 \|P_\Delta\| \\ \epsilon_1 \|P_E\| & \epsilon\epsilon_3 \|P_\Delta\| & \lambda_o - 2\epsilon_3 \|P_{\tilde{\Delta}}\| \end{bmatrix}, \\
\gamma^T &= \begin{bmatrix} 0 & \epsilon l_{\Delta_1} \|P_\Delta\| & 0 \end{bmatrix},
\end{aligned}
\tag{A.35}
$$

and $\lambda_e = \lambda_{min}(Q_E)$, $\lambda_\Delta = \lambda_{min}(Q_\Delta)$, $\lambda_o = \lambda_{min}(A_{\tilde{\delta}})$.

Now consider the matrix Λ. This matrix will be positive–definite if $\lambda_\Delta > l_{\Delta_2}$, $\lambda_o > 2\epsilon_3 \|P_{\tilde{\delta}}\|$ and the ϵ_i's are sufficiently small (which may be achieved by proper choice of the gain matrices). Note however that ϵ appears only in the off–diagonal terms of this matrix. A very small ϵ will increase the $\Lambda(2,1)$ element while a very large ϵ increases $\Lambda(2,3)$ and $\Lambda(2,1)$ elements. The element $\Lambda(2,1)$ is minimum for $\epsilon = \sqrt{\epsilon_0/(l_E\|P_\Delta\|)}$ which should typically be made small by design. As explained earlier, ϵ_0 depends on $K_\delta(q)$ and $K_{\tilde{\delta}}(q)$. On the other hand, the element $\Lambda(2,1)$ is directly affected by ϵ_0. Thus a reduction of ϵ_0 will be desirable since it will result in a smaller off diagonal term.

For $\epsilon = \sqrt{\epsilon_0/(l_E\|P_\Delta\|)}$, it is easy to obtain a condition on the elements of Λ so that it is positive–definite, i.e.,

$$\lambda_e(\lambda_\Delta - l_{\Delta_2} - 2\|P_\Delta\|\epsilon_2)(\lambda_o - 2\epsilon_3\|P_{\tilde{\delta}}\|) + 4\|P_E\|\|P_\Delta\|\epsilon_0\epsilon_3 >$$
$$\frac{\epsilon_0(\lambda_e\epsilon_3^2\|P_\Delta\|^2 + (\lambda_\Delta - l_{\Delta_2} - 2\|P_\Delta\|\epsilon_2)\|P_E\|^2)}{l_E\|P_\Delta\|} +$$
$$4(\lambda_o - 2\epsilon_3\|P_{\tilde{\delta}}\|)\epsilon_0 l_E\|P_\Delta\|. \tag{A.36}$$

It should also be noted that the effect of $\|K_d\|$ appears in the form of a product term with $\|P_E\|$ (in (A.35)), and since $\|Q_E\|$ is constant the variation of $\|K_d\|$ will not much affect the corresponding $\Lambda(1,3)$ term.

Now, suppose that for a given system and controller parameters, Λ is positive definite. Let us define the sets

$$\begin{aligned}
\mathcal{R} &= \{(E, \hat{\Delta}, \tilde{\delta}) \mid [\|E\| \ \|\hat{\Delta}\| \ \|\tilde{\delta}\|]\Lambda[\|E\| \ \|\hat{\Delta}\| \ \|\tilde{\delta}\|]^T \\
&\leq 2\gamma^T[\|E\| \ \|\hat{\Delta}\| \ \|\tilde{\delta}\|]^T\} \cap \Omega_i \\
\mathcal{S} &= \{(E, \hat{\Delta}, \tilde{\delta}) \mid V(E, \hat{\Delta}, \tilde{\delta}) \leq c_1\} \subset \Omega_i \\
\mathcal{T} &= \{(E, \hat{\Delta}, \tilde{\delta}) \mid V(E, \hat{\Delta}, \tilde{\delta}) = c_2\} \subset \Omega_i \tag{A.37}
\end{aligned}$$

where $0 < c_2 \leq c_1$ and c_2 is the smallest constant such that $\mathcal{R} \subset \mathcal{T}$. Since the trajectory defined by y_r, $\dot{y}_r$, $\ddot{y}_r$ is bounded, and Λ is positive–definite, $\mathcal{R}$ is uniformly bounded. If the initial state is outside $\mathcal{S} - \mathcal{R}$, where $\dot{V} \leq 0$, it follows that there exists a finite time t_f such that any solution starting from $\mathcal{S} - \mathcal{R}$, at $t > 0$, will enter $\mathcal{T}$ at t_f, and remain in $\mathcal{T}$ thereafter for all $t \geq t_f$. The residual set $\mathcal{T}$ encompasses an ellipsoid in $[\|E\| \ \|\hat{\Delta}\| \ \|\tilde{\delta}\|]$ coordinates with its bounds being $\sqrt{\gamma^T\Lambda^{-1}\gamma/\lambda_{min}(\Lambda)}$ and $\sqrt{\gamma^T\Lambda^{-1}\gamma/\lambda_{max}(\Lambda)}$ which are both of order ϵ. Since $\hat{\Delta} = \epsilon\Delta$, it follows that Δ is of the order of $\frac{1}{\epsilon}$ on this residual set. As discussed earlier, controller parameters should be chosen such that Λ is positive–definite. To this end, some qualitative robustness measures may be obtained. In particular, making the norms of P_E, $K_\delta(q)$, $K_{\dot{\delta}}(q)$, $P_{\dot{\delta}}$, P_Δ small yields a better margin of robustness.

A.4. Proof of Theorem 5.1

Consider the dynamic equations of the closed–loop system given by (5.13), (5.16) and (5.20) which are repeated here for convenience

$$\dot{s}_{\sigma_i} = (1 + \Delta B_{ii}^*)k_i sat(\gamma_i s_{\sigma_i}) + \sum_{j=1, j\neq i}^{n} \Delta B_{ij}^* k_j sat(\gamma_j s_{\sigma_j}) + \eta_i^*,$$

$$i = 1, \cdots, n \tag{A.38}$$

$$\dot{E} = A_E E + d_E(E, \Delta, s_\sigma, t) \tag{A.39}$$

$$\dot{\Delta} = A_\Delta(q)\Delta + g_\Delta(E, \Delta, s_\sigma, t). \tag{A.40}$$

Since A_E and $A_\Delta(q)$ are Hurwitz matrices the following Lyapunov equations are satisfied

$$\begin{aligned} A_E^T P_E + P_E A_E &= -Q_E \\ A_\Delta^T(q)P_\Delta(q) + P_\Delta(q)A_\Delta(q) &= -Q_\Delta \end{aligned} \tag{A.41}$$

where P_E, Q_E, $P_\Delta(q)$ and Q_Δ are symmetric positive–definite matrices. Let us choose the positive–definite Lyapunov function candidate

$$V = E^T P_E E + \hat{\Delta}^2 P_\Delta(q)\hat{\Delta} + 0.5 s_\sigma^T s_\sigma \tag{A.42}$$

where $\hat{\Delta} = \epsilon_\Delta \Delta$ with ϵ_Δ being a small positive scaling constant (typically less than one). The need for this scaling factor arises from the stability requirement that E and s_σ converge to small values near zero while Δ remains bounded. Thus, the scaling factor allows us to show that E, s_σ and $\hat{\Delta}$ converge to small values, hence Δ converges to $\hat{\Delta}/\epsilon_\Delta$.

Taking the time derivative of V yields

$$\dot{V} = -E^T Q_E E - \hat{\Delta}^T Q_\Delta \hat{\Delta} + 2 d_E^T P_E E + 2 \epsilon_\Delta \hat{\Delta} P_\Delta g_\Delta + \hat{\Delta} \dot{P}_\Delta \hat{\Delta} + \dot{s}_\sigma s_\sigma.$$
$$(A.43)$$

Consider a continuous bounded reference trajectory (at least C^2) and a bounded region Ω_1 containing the origin of (E, Δ, s_σ). Then for all

$$(E, \Delta, s_\sigma) \in \Omega_1 \subset \mathbf{R}^{2m+3n}$$

with $\|s_\sigma\|_1 < 1$ as described later, we have

$$\|d_E\| \leq l_1 + l_2 \|\Delta\| + l_3 \|E\| \qquad (A.44)$$

where, as far as the design parameters are concerned, l_1 is affected by K and the reference trajectory, l_2 is affected by K_δ and $K_{\dot{\delta}}$, and l_3 by K_p and K_d. Note that all the norms in the previous inequality and subsequent discussion are 1–norms. Similarly for all $(E, \Delta, s_\sigma) \in \Omega_2 \subset \mathbf{R}^{2m+3n}$

$$\|g_\Delta\| \leq l_4 + l_5 \|\Delta\| + l_6 \|E\| \qquad (A.45)$$

with l_4 affected by K and the reference trajectory, and l_6 by K_p and K_d.

A detailed analysis also reveals that for all $(E, \Delta, s_\sigma) \in \Omega_3 \subset R^{2m+3n}$, where Ω_3 is a finite region containing the origin of (E, Δ, s_σ), we have $\|\dot{P}_\Delta(q)\| \leq l_{\Delta_2}$. This can be shown by noting that $P_\Delta(q)$ has a finite growth rate with respect to q (see e.g. [67], Problem 5.21) and the reference trajectory is bounded.

As discussed previously the surface $s_\sigma = 0$ is attractive for $s_\sigma \in R$ and other state variables in a closed set Ω^* defined earlier. Thus for any $s_\sigma \in R$, there is a finite time after which s_σ will be small enough such that $\|s_\sigma\| < 1$ (or $\|s_\sigma\| > \|s_\sigma\|^2$). This time can be made short by increasing β_i's in section A.2.

Hence, letting $\Omega_i = \Omega_1 \cap \Omega_2 \cap \Omega_3$ and substituting the previous inequalities in $\dot{V}$ and rearranging the right–hand side of the resulting inequality in terms of $\|E\|$, $\|\Delta\|$ and $\|s_\sigma\|$ will, after some algebraic manipulations, yield

$$\dot{V} \leq -[\|E\|\ \|\hat{\Delta}\|\ \|s_\sigma\|]\Lambda \begin{bmatrix} \|E\| \\ \|\hat{\Delta}\| \\ \|s_\sigma\| \end{bmatrix} + 2\gamma^T \begin{bmatrix} \|E\| \\ \|\hat{\Delta}\| \\ \|s_\sigma\| \end{bmatrix} \quad \forall (E, \Delta, s_\sigma) \in \Omega_i$$
$$(A.46)$$

where

$$\Lambda = \begin{bmatrix} \lambda_e - 2l_3\|P_E\| & \frac{l_2\|P_E\|}{\epsilon_\Delta} + l_6\epsilon_\Delta\|P_\Delta\| & 0 \\ \frac{l_2\|P_E\|}{\epsilon_\Delta} + \epsilon_\Delta l_6\|P_\Delta\| & \lambda_\Delta - l_7 - 2\|P_\Delta\|l_5 & 0 \\ 0 & 0 & \beta_{min} \end{bmatrix},$$

$$\gamma^T = [l_1\|P_E\| \quad \epsilon_\Delta l_4\|P_\Delta\| \quad 0], \tag{A.47}$$

where $\beta_{min} = min(\beta_1, \cdots, \beta_n)$ and $\lambda_e = \lambda_{min}(Q_E)$, $\lambda_\Delta = \lambda_{min}(Q_\Delta)$, $\lambda_o = \lambda_{min}(A_{\tilde{\delta}})$.

Now consider the matrix Λ when no uncertainty is present. In such a case it is positive–definite if $\lambda_\Delta > l_7 + 2l_5\|P_\Delta\|$ and $\lambda_e(\lambda_\Delta - l_7 - 2l_5\|P_\Delta\|) > (\frac{l_2\|P_E\|}{\epsilon_\Delta} + l_6\epsilon_\Delta\|P_\Delta\|)^2$. Thus a neighborhood of the case with no uncertainty can be found such that Λ is positive–definite. Let us define the sets

$$\begin{aligned} \mathcal{R} &= \{(E, \hat{\Delta}, s_\sigma) \mid [\|E\| \; \|\hat{\Delta}\| \; \|s_\sigma\|]\Lambda[\|E\| \; \|\hat{\Delta}\| \; \|s_\sigma\|]^T \\ &\leq 2\gamma^T[\|E\| \; \|\hat{\Delta}\| \; \|s_\sigma\|]^T\} \cap \Omega_i \\ \mathcal{S} &= \{(E, \hat{\Delta}, s_\sigma \mid V(E, \hat{\Delta}, s_\sigma) \leq c_1\} \subset \Omega_i \\ \mathcal{T} &= \{(E, \hat{\Delta}, s_\sigma) \mid V(E, \hat{\Delta}, s_\sigma) = c_2\} \subset \Omega_i \end{aligned} \tag{A.48}$$

where $0 < c_2 \leq c_1$ and c_2 is the smallest constant such that $\mathcal{R} \subset \mathcal{T}$. Since the trajectory defined by $y_r, \dot{y}_r, \ddot{y}_r$ is bounded, and Λ is positive–definite, $\mathcal{R}$ is uniformly bounded. If the initial state is outside $\mathcal{S} - \mathcal{R}$, where $\dot{V} \leq 0$, it follows that there exists a finite time t_f such that any solution starting from $\mathcal{S} - \mathcal{R}$, at $t > 0$, will enter $\mathcal{T}$ at t_f, and remain in $\mathcal{T}$ thereafter for all $t \geq t_f$. The residual set $\mathcal{T}$ encompasses an ellipsoid in $[\|E\| \; \|\hat{\Delta}\| \; \|\tilde{\delta}\|]$ coordinates with its bounds being $\sqrt{\gamma^T\Lambda^{-1}\gamma/\lambda_{min}(\Lambda)}$ and $\sqrt{\gamma^T\Lambda^{-1}\gamma/\lambda_{max}(\Lambda)}$ which are both of order ϵ. Since $\hat{\Delta} = \epsilon_\Delta\Delta$, it follows that Δ is of the order of $\frac{1}{\epsilon_\Delta}$ on this residual set. As discussed earlier, controller parameters should be chosen such that Λ is positive–definite. By considering the terms affected by these gains some qualitative robustness measures may be obtained. In particular, making the norms of P_E, $K_\delta(q)$, $K_{\dot{\delta}}(q)$, $P_{\tilde{\delta}}$, P_Δ small yields a better margin of robustness.

Appendix B.

Kinematic Description

The various terms used in the manipulators considered in this paper are illustrated here for a two–link planar manipulator as shown in Figure B.1. Assuming that the deformation of each link is small compared to its length, the length of each deformed link is approximately equal to the length of the line–segment joining the two ends of the link. Axis X_2 is defined to be the tangent line drawn from the tip of the first link. Thus, angles y_1 and y_2 are defined as $y_1 = q_1 + \sum_i \phi_{1i}(l_1)\delta_{1i}/l_1$ and $y_2 = q_2 + \sum_j \phi_{2j}(l_2)\delta_{2j}/l_2$. where ϕ_{1i} and ϕ_{2j} represent the modal shape functions of the links (clamped shape functions have been assumed). The control laws derived in this paper result in small tracking errors for y_1 and y_2 defined above. However, if a Cartesian trajectory is specified, it is convenient to define another output for y_2 as follows: Extending the line AB, let us define y_{2n} as the angle between lines BD and BC. Thus, the *new* y_2 can be written as $y_{2n} = y_2 - y_1 + q_1 + \Psi_1$. The angle Ψ_1 is the slope of the tangent at B and is given by $\Psi_1 = \sum_i \phi'_{1ie}\delta_{1i}$, where ϕ'_{1ie} is the spatial derivative of the i–th shape function at l_1. Thus we have $y_{2n} = q_2 + \sum_j \phi_{2j}(l_2)\delta_{2j}/l_2 + \sum_i (\phi'_{1ie} - \phi_{1i}(l_1)/l_1)\delta_{1i}$ which together with y_1 may be represented in the form given by (2.8).

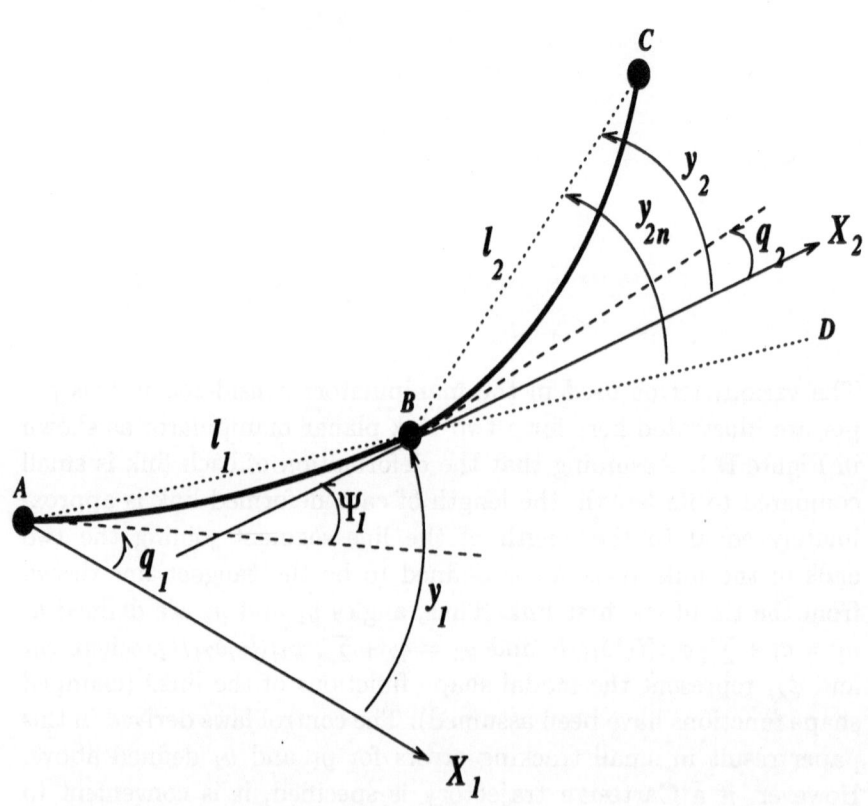

Figure B.1. Kinematic description for a flexible two–link manipulator.

Appendix C.
Dynamic Models

C.1. Dynamic Model of the Flexible Single–link Arm

The dynamic model used in designing the controller for the experimental setup is derived based on the assumed modes method with clamped–mass shape functions given by

$$\phi_i(x) = cosh(\lambda_i x/l) - cos(\lambda_i x/l) - \gamma_i(sinh(\lambda_i x/l) - sin(\lambda_i x/l)) \tag{C.1}$$

where l is the length of the link, x is the position variable along the link and λ_i's are obtained from

$$1 + cosh(\lambda_i)cos(\lambda_i) + \frac{M_p}{m}\lambda_i(sinh(\lambda_i)cos(\lambda_i) - cosh(\lambda_i)sin(\lambda_i))$$
$$= 0 \tag{C.2}$$

where $m = 0.210kg$ is the mass of the link, and $M_p = 0.251kg$ is the payload mass. The first three λ_i's are: 1.2030, 4.0159, and 7.1243.

The mass and stiffness matrices and Coriolis and centrifugal terms obtained by using $MAPLE$ [35] are as follows (see Section 2 for the definition of the terms)

$$M(q,\delta) = \begin{bmatrix} m(\delta) & 0.1863 & 0.0208 \\ 0.1863 & 0.2655 & -7.1518 \times 10^{-5} \\ 0.0208 & -7.1518 \times 10^{-5} & 0.2162 \end{bmatrix},$$

$$K = \begin{bmatrix} 7.5340 & -0.0004 \\ -0.0004 & 755.5287 \end{bmatrix},$$

$$\begin{aligned}
g^T(q,\dot{q},\delta,\dot{\delta}) = \ & [\dot{q}\dot{\delta}_1(0.2667\delta_1 + 4.2261 \times 10^{-4}\delta_2) \\
+ \ & \dot{q}\dot{\delta}_2(0.5308\delta_2 + 4.2261 \times 10^{-4}\delta_1), \\
- \ & 0.5\dot{q}^2(0.2667\delta_1 + 4.2261 \times 10^{-4}\delta_2), \quad\quad\text{(C.3)} \\
- \ & 0.5\dot{q}^2(0.5308\delta_2 + 4.2261 \times 10^{-4}\delta_1)]. \quad\quad\text{(C.4)}
\end{aligned}$$

where $m(\delta) = 0.1334 + 0.2654\delta_1^2 + 0.2149\delta_2^2 + 4.2261 \times 10^{-4}\delta_1\delta_2$ and

$$g(q,\dot{q},\delta,\dot{\delta}) = [g_1^T \ g_2^T]^T.$$

The natural frequencies obtained from this model can be derived as

$$\frac{1}{2\pi}\sqrt{eig(H_{220}K)} = 5.2059, \ 21.7267Hz$$

which are close to the experimental values 5.5 and 20 Hz.

C.2.   Dynamic Model of the Flexible Two–link Arm

The dynamic model used in designing the controller for the experimental setup is derived based on the assumed modes method with clamped–mass shape functions given by

$$\begin{aligned}
\phi_i(\sigma) = \ & cosh(\lambda_i\sigma/l_2) - cos(\lambda_i\sigma/l_2) \\
- \ & \gamma_i(sinh(\lambda_i\sigma/l_2) - sin(\lambda_i\sigma/l_2)) \quad\quad\text{(C.5)}
\end{aligned}$$

where l_2 is the length of the flexible link, σ is the position variable along the link, and the λ_i's are obtained from

$$\begin{aligned}
1 \ + \ & cosh(\lambda_i)cos(\lambda_i) \\
+ \ & \frac{M_p}{m}\lambda_i(sinh(\lambda_i)cos(\lambda_i) - cosh(\lambda_i)sin(\lambda_i)) = 0 \quad\text{(C.6)}
\end{aligned}$$

where $m = 0.210kg$ is the mass of the second link , and $M_p = 0.251kg$ is the payload mass. The first three λ_i's are: 1.2030, 4.0159, and 7.1243.

The elements of the mass and stiffness matrices and Coriolis and centrifugal terms obtained using $MAPLE$ [35] for one flexible mode are as follows

$$
\begin{aligned}
M(1,1) &= m_1 + m_2 cos(q_2) + m_3\delta^2 \\
M(1,2) &= M(2,1) = m_4 + m_5 cos(q_2) + m_6\delta^2 \\
M(1,3) &= M(3,1) = m_7 + m_8 cos(q_2) \\
M(2,2) &= m_9 + m_{10}\delta^2 \\
M(2,3) &= M(3,2) = m_{11} \\
M(3,3) &= m_{12} \\
f_1(1) + g_1(1) &= \dot{\delta}sin(q_2)[(m_{14} - m_8)\dot{q}_2 + m_{13}\dot{q}_1 \\
&\quad - (m_5\dot{q}_2^2 + m_2\dot{q}_2\dot{q}_1)] \\
f_1(2) + g_1(2) &= sin(q_2)[\dot{\delta}(m_8 + m_{14})\dot{q}_1 - (m_5\dot{q}_2^2 + m_2\dot{q}_2\dot{q}_1)] \\
f_2(1) + g_2(1) &= -sin(q_2)[m_8\dot{q}_1 + 0.5m_{13}\dot{q}_1^2 + m_{14}\dot{q}_2\dot{q}_1] \quad \text{(C.7)}
\end{aligned}
$$

where $M(i, j)$ represents the (i, j)th element of the mass matrix and $f_i(j) + g_i(j)$ represents element j of the i-th Coriolis and centrifugal terms.

The numerical values of the parameters in (C.7) are given below in appropriate SI units, i.e.,

$$
\begin{aligned}
m_1 &= 0.2255, \quad m_2 = 0.1090, \quad m_3 = 0.2654, \quad m_4 = 0.1332, \\
m_5 &= 0.5453, \quad m_6 = 0.2654, \quad m_7 = 0.1862, \quad m_8 = 0.0727, \\
m_9 &= 0.1332, \quad m_{10} = 0.2654, \quad m_{11} = 0.1862, \quad m_{12} = 0.2654, \\
m_{13} &= -0.14542, \quad m_{14} = -0.7271, \quad K = 7.5340.
\end{aligned}
$$

$$\text{(C.8)}$$

C.3. Derivation of the Dynamic Equations

Consider the manipulator sketched in Figure C.1. Assume that link 1 has cross sectional area $A_1(x_1)$, length l_1, with flexible modes represented by $\delta_{11}, \delta_{12}, \cdots, \delta_{1m_1}$. Similarly, link 2 has cross sectional area $A_2(x_2)$, length l_2, and flexible modes $\delta_{21}, \delta_{22}, \cdots, \delta_{2m_2}$, where $m_1 + m_2 = m$ is the total number of flexible modes. The $X - Y$ frame is the stationary $world$ frame, q_1 is the angle of the tangent at $x_1 = 0$ with respect to the X–axis, and q_2 is the angle that x_2–axis makes relative to the slope of the end point of link 1. The modal

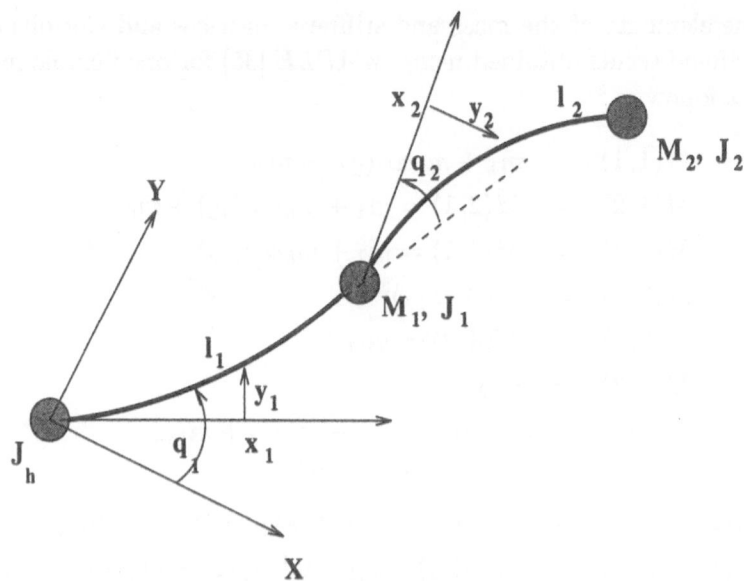

Figure C.1. A two–link flexible manipulator.

shape functions are assumed to be *clamped* at the actuation end (clamped–mass or clamped–free).

Using the above kinematic description the dynamic equations can then be obtained by utilizing the Lagrangian formulation. Towards this end, first the kinetic energy of the system is obtained. Let us denote the description of a point on link 1, written in the $X - Y$ plane, by $r_1(x_1, t)$. Then

$$r_1(x_1, t) = T_1(q_1) \begin{bmatrix} x_1 \\ \sum_{i=1}^{m_1} \phi_{1i}(x_1)\delta_{1i} \end{bmatrix} \tag{C.9}$$

where $T_1(q_1)$ is the rotation matrix of the x_1–y_1 frame, i.e.,

$$T_1(q_1) = \begin{bmatrix} cosq_1 & -sinq_1 \\ sinq_1 & cosq_1 \end{bmatrix}. \tag{C.10}$$

Therefore the kinetic energy due to link 1 is given by

$$T_{k_1} = \frac{1}{2} \int_0^{l_1} \rho A_1(x_1) \dot{r}_1^T \dot{r}_1 dx_1. \tag{C.11}$$

Similarly the description of a point corresponding to x_2 on the second link is given in the X–Y frame by

$$r_1(x_1, t) = T_1(q_1)(\begin{bmatrix} l_1 \\ y_1(l_1) \end{bmatrix} + T_2 \begin{bmatrix} x_2 \\ y_2 \end{bmatrix}) \tag{C.12}$$

where

$$y(l_1) = \sum_{i=1}^{m_1} \phi_{1i}(l_1)\delta_{1i}$$

$$T_2 = \left[\begin{array}{cc} cos(q_2 + \sum_{i=1}^{m_1} \phi'_{1ie}\delta_{1i}) & -sin(q_2 + \sum_{i=1}^{m_1} \phi'_{1ie}\delta_{1i}) \\ sin(q_2 + \sum_{i=1}^{m_1} \phi'_{1ie}\delta_{1i}) & cos(q_2 + \sum_{i=1}^{m_1} \phi'_{1ie}\delta_{1i}) \end{array} \right]$$

$$\phi'_{1ie} := \frac{d}{dx_1}(\phi_{1i}(x_1))\mid_{x_1=l_1}. \tag{C.13}$$

Hence the kinetic energy due to the second link is

$$T_{k_2} = \frac{1}{2}\int_0^{l_2} \rho A_2(x_2)\dot{r}_2^T \dot{r}_2 dx_2. \tag{C.14}$$

The hub kinetic energy is given by

$$T_h = \frac{1}{2}J_h\dot{q}_h^2 \tag{C.15}$$

where

$$q_h = q_1 + \sum_{i=1}^{m_1} \phi'_{1i}(0)\delta_{1i}. \tag{C.16}$$

Since $\phi'_{1i}(0) = 0$ for clamped mode shapes, we have

$$T_h = \frac{1}{2}J_h\dot{q}_1^2. \tag{C.17}$$

The kinetic energy due to M_1 and J_1, which denote the mass and mass moment of inertia of the case and stator of the second motor, is given by

$$T_{k_{mot}} = \frac{1}{2}M_1\dot{r}_1^T\dot{r}_1\mid_{x_1=l_1} +\frac{1}{2}J_1(\dot{q}_1 + \sum_{i=1}^{m_1} \phi'_{1ie}\dot{\delta}_{1i})^2. \tag{C.18}$$

And finally the payload kinetic energy is

$$T_{k_p} = \frac{1}{2}M_2\dot{r}_2^T\dot{r}_2\mid_{l_2} +\frac{1}{2}J_2(\dot{q}_1 + \dot{q}_2 + \sum_{i=1}^{m_1} \phi'_{1ie}\dot{\delta}_{1i} + \sum_{i=1}^{m_2} \phi'_{2ie}\dot{\delta}_{2i})^2 \tag{C.19}$$

This the total kinetic energy is obtained as

$$T_k = T_{k_1} + T_{k_2} + T_h + T_{k_{mot}} + T_{k_p}. \tag{C.20}$$

The elastic potential energy is obtained from

$$V_e = \frac{1}{2} \sum_{i=1}^{m_1} \sum_{l=1}^{m_1} \delta_{1k} \delta_{1l} k_{1kl} + \frac{1}{2} \sum_{k=1}^{m_2} \sum_{l=1}^{m_2} \delta_{2k} \delta_{2l} k_{2kl} \qquad \text{(C.21)}$$

which can be written in the matrix form $V_e = \frac{1}{2} \delta^T K \delta$ and

$$k_{1kl} = \int_0^{l_1} EI_{z1}(x_1) \frac{d^2 \phi_{1k}(x_1)}{dx_1^2} \frac{d^2 \phi_{1l}(x_1)}{dx_1^2} dx_1$$

$$k_{2kl} = \int_0^{l_2} EI_{z2}(x_2) \frac{d^2 \phi_{2k}(x_2)}{dx_2^2} \frac{d^2 \phi_{2l}(x_2)}{dx_2^2} dx_2 \qquad \text{(C.22)}$$

where E is the modulus of elasticity of the material and $I_{z1}(x_1)$, $I_{z2}(x_2)$ are the area moments of inertia about the axis of rotation z_1, z_2 at x_1 and x_2. Note that for a uniform manipulator the cross product terms in (C.22) are zero if orthogonal shape functions (e.g., clamped–free) are used. If gravity is present, the potential energy due to gravity can also be added to V_e. However, we derive the dynamic equations in the absence of gravity. Denoting the degrees of freedom by

$$z^T = [q_1 \ q_2 \ \delta_{11} \ \delta_{12} \ \cdots \ \delta_{1m_1} \ \delta_{21} \ \delta_{22} \ \cdots \ \delta_{2m_2}] := [q^T \ \delta^T] \quad \text{(C.23)}$$

the Lagrangian equation for the system is given by

$$L = T_k - V_e = \frac{1}{2}(\dot{z}^T M \dot{z} - z^T K z) \qquad \text{(C.24)}$$

from which the system dynamics can be obtained using

$$\frac{d}{dt} \frac{\partial L}{\partial \dot{z}} - \frac{\partial L}{\partial z} = \tau_g \qquad \text{(C.25)}$$

where τ_g is the generalized vector of torques which, for the case of clamped mode shapes is given by

$$\tau_g = \begin{bmatrix} I_{2\times2} \\ 0_{2\times2} \end{bmatrix} \tau := Q\tau \qquad \text{(C.26)}$$

where τ is the actual torques at the joints.

The mass matrix of the manipulator can thus be obtained by writing the kinetic energy given by (C.21) and putting it in the form $\frac{1}{2}\dot{z}^T M(q, \delta)\dot{z}$. Then, using the Lagrangian (C.25), it follows that

$$M(q, \delta)\ddot{z} + f(z, \dot{z}) + K_a z = Q\tau \qquad \text{(C.27)}$$

where the i–th element of $f(z, \dot{z})$ is

$$f_i(z, \dot{z}) = e_i^T \sum_{j=1}^{m+2} \dot{z}_j \frac{\partial M}{\partial z_j} \dot{z} - \frac{1}{2} \dot{z}^T \frac{\partial M}{\partial z_i} \dot{z}, \quad i = 1, \cdots, m+2 \quad \text{(C.28)}$$

with e_i the unity vector [1], $f(z, \dot{z})$ containing the centrifugal and Coriolis terms, and K_a is given by

$$K_a = \begin{bmatrix} 0_{2\times 2} & 0_{2\times m} \\ 0_{m\times 2} & K_{m\times m} \end{bmatrix}. \quad \text{(C.29)}$$

The above steps were coded to obtain the dynamic equations using the symbolic manipulation software *MAPLE* [35].

---

[1]A unity vector e_i is a vector with all its elements zero except for the i–th element; which is 1.

References

[1] G. Hastings and W. J. Book, "Experiments in the Optimal Control of a Flexible Link Manipulator," *American Control Conference*, pp. 728–729, Boston, 1985.

[2] R. H. Cannon and E. Schmitz, "Initial Experiments on the End–Point control of a Flexible One–Link Robot," *International Journal of Robotics Research*, Vol. 3, No. 3, pp. 62–75, 1984.

[3] G. Hastings and W. J. Book, "A Linear Dynamic Model for Flexible Robotic Manipulators," *IEEE Control Systems Magazine*," pp. 61–64, 1987.

[4] F. Bellezza, L. Lanari and G. Ulivi, "Exact Modeling of the Flexible Slewing Link," *IEEE International Conference on Robotics and Automation*, pp. 734–739, 1990.

[5] S. Cetinkunt and W. L. Yu, "Closed Loop Behavior of a Feedback–Controlled Flexible Arm: A Comparative Study," *International Journal of Robotics Research*, Vol. 10, No. 3, pp. 263–275, 1991.

[6] E. Bayo, "A Finite–Element Approach to Control the End–Point Motion of a Single–link Flexible Robot," *Journal of Robotic Systems*, Vol. 4, No. 1, pp. 63–75, 1987.

[7] P. B. Usoro, R. Nadira and S. S. Mahil, "A Finite Element/Lagrange Approach to Modeling Lightweight Flexible Manip-

ulators,"*International Journal of Robotics Research*, Vol. 3, No. 3, pp. 87–101, 1984.

[8]   B. Siciliano and W. J. Book, "A Singular Perturbation Approach to Control of Lightweight Flexible Manipulators," *International Journal of Robotics Research*, Vol. 7, No. 4, pp. 79–90, 1989.

[9]   A. De Luca A. and B. Siciliano, "Explicit Dynamic Modeling of a Planar Two–link Flexible Manipulator," *29th IEEE Conference on Decision and Control*, Honolulu, Hawaii, December 1990.

[10]   M. Benati and A. Morro, "Dynamics of Chain of Flexible Links," *ASME Journal of Dynamic Systems, Measurement, and Control*, Vol. 110, pp. 410–415, 1988.

[11]   H. P. Pota and M. Vidyasagar, "Passivity of Flexible Beam Transfer Functions with Modified Outputs," *IEEE International Conference on Robotics and Automation*, Sacramento, CA, pp. 2826–2831, April 1991.

[12]   D. Wang and M. Vidyasagar, "Passive Control of a Flexible Link," *International Journal of Robotics Research*, Vol. 11, No. 6, pp. 572–578, Dec. 1992.

[13]   T. Yoshikawa, H. Murakami and K. Hosoda, "Modeling and Control of a 3 D.O.F. Manipulator with two Flexible Links," *29th IEEE Conference on Decision and Control*, Honolulu, Hawaii, pp. 2532–2537, December 1990.

[14]   A. De luca, L. Lanari, P. Lucibello, S. Panzieri and G. Ulivi, "Control Experiments on a Two Link Robot with a Flexible Forearm," *29th IEEE Conference on Decision and Control*, Honolulu, Hawaii, pp. 520–527 December 1990.

[15]   L. Meirovitch, *Elements of Vibration Analysis*, McGraw–Hill, New-York, 1975.

[16]   Y. Aoustin and C. Chevallereau, "The Singular Perturbation Control of a Flexible-link Robot," *IEEE International Conference on Robotics and Automation*, Atlanta, GA, pp. 737–742, May 1993.

[17]   B. Siciliano, V. R. J. Prasad and A. J. Calise, "Design of a Composite Controller for a Two–link Flexible Manipulator," *International Symposium on Intelligent Robotics*, Bangalore, India, pp. 126–137, 1991.

[18]   R. M. Hirschorn, "Invertibility of Nonlinear Control Systems," *SIAM Journal of Control and Optimization*, Vol. 17, No. 2, pp. 282–297, 1979.

[19] C. I. Byrnes and A. Isidori, "Global Feedback Stabilization of Non-linear Systems," *24th IEEE Conference on Decision and Control*, pp. 1031–1037, 1985.

[20] L. Russell and F. W. Wang, "Minimum–weight Robot Arm for a Specified Fundamental Frequency," *IEEE International Conference on Robotics and Automation*, pp. 490-495, Atlanta, GA, 1993.

[21] H. Asada, J. H. Park and S. Rai, "A Control–Configured Flexible Arm: Integrated Structure–Control Design," *IEEE International Conference on Robotics and Automation*, Sacramento, CA, 1991.

[22] J. H. Park and H. Asada, "Integrated Structure / Control Design of a Two–link Nonrigid Robot Arm for High Speed Positioning," *IEEE International Conference on Robotics and Automation*, Nice, France, 1992.

[23] A. De Luca and L. Lanari, "Achieving Minimum Phase Behavior in a One–Link Flexible Arm," *International Symposium on Intelligent Robotics*, Bangalore, India, 1991.

[24] D. S. Kwon and W. J. Book, "An Inverse Dynamic Method Yielding Flexible Manipulator State Trajectories," *American Control Conference*, pp. 186–193, 1990.

[25] K. L. Hillsley and S. Yurkovich, "Vibration Control of a Two–link Flexible Robot Arm," *International Conference on Robotics and Automation*, Sacramento, CA, pp. 2121–2126, 1991.

[26] A. Tzes and S. Yurkovich, "An Adaptive Input Shaping Control Scheme for Vibration Suppression in Slewing Flexible Structures," *IEEE Transactions on Control Systems Technology*, Vol.1, No. 2, pp. 114–121, June 1993.

[27] F. Khorrami, S. Jain and A. Tzes, "Experiments on Rigid Body Based Controllers with Input Preshaping for a Two Link flexible Manipulator," *American Control Conference*, pp. 2957–2961, 1992.

[28] S. E. Burke and E. Hubbard, "Distributed Actuator Control Design for Flexible Beams," *Automatica*, Vol. 24, No. 5, pp. 619–627, 1988.

[29] W. J. Book, "Recursive Lagrangian Dynamics of Flexible Manipulator Arms," *International Journal of Robotics Research*, Vol. 3, No. 3, pp. 87–101, Fall 1984.

[30] K. Khorasani, " A Robust Adaptive Control Design for a Class of Dynamical Systems using Corrected Models," *IEEE Transactions on Automatic Control*, Vol.39, No. 6, pp. 1726–1732, 1994.

[31]   A. Isidori, *Nonlinear Control Systems*, Springer–Verlag, New York, 1995.

[32]   M. Spong and M. Vidyasagar, *Robot Dynamics and Control*, J. Wiley ans Sons, New York, 1989.

[33]   W. T. Thomson, *Theory of Vibrations with Applications*, Prentice–Hall, Englewood Cliffs, NJ, 1988.

[34]   S. Tosunoglu, S. H. Lin and D. Tesar, "Accessibility and Controllability of Flexible Robotic Manipulators," *ASME Journal of Dynamic Systems, Measurement, and Control*, Vol. 114, pp. 50–58, 1992.

[35]   D. Redfern, *The Maple Handbook*, Springer-Verlag, New York, 1993.

[36]   D. Wang and M. Vidyasagar, "Transfer Functions for a Single Flexible Link," *IEEE International Conference on Robotics and Automation*, pp. 1042–1047, 1989.

[37]   A. De Luca and L. Lanari, "Achieving Minimum–phase Behavior in a One–Link Flexible Arm," *International Symposium on Intelligent Robotics*, Bangalore, India, pp. 224–235, 1991.

[38]   S. K. Madhavan and S. N. Singh, "Inverse Trajectory Control and Zero–Dynamics Sensitivity of an Elastic Manipulator," *International Journal of Robotics and Automation*, Vol. 6, No. 4, pp. 179–191, 1991.

[39]   V.G. Moudgal, K. Passino and S. Yurkovich, "Rule–Based Control for a Flexible–link Robot," *IEEE Transactions on Control Systems Technology*, Vol. 2, No. 4, pp. 392–405, Dec. 1994.

[40]   E. Bayo and H. Moulin, "An Efficient Computation of the Inverse Dynamics of Flexible Manipulators in the Time Domain," *IEEE International Conference on Robotics and Automation*, pp. 710–715, 1989.

[41]   R.V. Patel and P. Misra, "Transmission Zero Assignment in Linear Multivariable Systems, Part II: The General Case," *American Control Conference*, pp. 644–648, Chicago, IL, 1992.

[43]   K. Hashtrudi–Zaad and K. Khorasani, "Control of Nonminimum Phase Singularly Perturbed Systems with Applications to Flexible Link Manipulators," *International Journal of Control*, Vol. 63, No. 4, pp. 679–701, March 1996.

[44]   J. H. Chow and P. V. Kokotović, "Two Time Scale Feedback Design of a Class of Nonlinear Systems," *IEEE Transactions on Automatic Control*, Vol. AC–23, pp. 438–443, 1978.

[45]   M. W. Spong, K. Khorasani and P. V. Kokotović, "An Integral Manifold Approach to the Feedback Control of Flexible Joint Robots," *IEEE Journal of Robotics and Automation*, Vol. 3, pp. 291–300, 1987.

[46]   K. Khorasani and M. W. Spong, "Invariant Manifolds and their Application to Robot Manipulators with Flexible Joints," *IEEE International Conference on Robotics and Automation*, pp. 978–983, 1985.

[47]   H. K. Khalil, "On the Robustness of Output Feedback Control Methods to Modeling Errors," *IEEE Transactions on Automatic Control*, Vol. AC–26, pp. 524–526, 1981.

[48]   F. L. Lewis and M. Vandegrift, "Flexible Robot Arm Control by a Feedback Linearization Singular Perturbation Approach," *IEEE International Conference on Robotics and Automation*, pp. 729–736, 1993.

[49]   B. Siciliano, J. V. R. Prasad and A. J. Calise, "Output Feedback Two–time Scale Control of Multi–link Flexible Arms," *ASME J. Dyn. Sys., Meas., Contr.*, Vol. 114, pp. 70–77, 1992.

[50]   Y. Aoustin, C. Chevallereau, A. Glumineau, and C. H. Moog, "Experimental Results for the End–Effector Control of a Single Flexible Robotic Arm," *IEEE Transactions on Control Systems Technology*, Vol. 2, pp. 371–381, 1994.

[51]   F. Khorrami, "A Two–stage Controller for Vibration Suppression of Flexible–link Manipulators," *29th IEEE Conference on Decision and Control*, pp. 2560–2565, 1990.

[52]   B. Siciliano, W. J. Book and G. De Maria, "An Integral Manifold Approach to Control of a One–Link Flexible arm," *25th IEEE Conference on Decision and Control*, pp. 1131–1134, 1986.

[53]   K. Khorasani," Adaptive Control of Flexible–joint Robots," *IEEE Transactions on Robotics and Automation*, Vol. 8, No. 2, pp. 250–267, 1992.

[54]   H. Geniele, R. V. Patel and K. Khorasani, "End–Point Control of a Flexible–link Manipulator: Theory and Experiment," *IEEE Trans. Control Systems Technology*, Vol. 5, pp 556–570, 1997.

[55]   J. Carusone, S. B. Keir and D'Eleuterio, "Experiments in End–Effector Tracking Control for Structurally Flexible Space Manipulators," *IEEE Transactions on Robotics and Automation*, Vol. 9, No. 5, pp. 553–560, October 1993.

[56] D. Wang and M. Vidyasagar, "Feedback Linearizability of Multi-link Manipulators with One Flexible Link," *28th IEEE Conference on Decision and Control*, Tampa, Florida, 1989.

[57] D. N. Schoenwald and and U. Özgüner, "On combining Slewing and Vibration Control in Flexible Manipulators via Singular Perturbations," *29th IEEE Conference on Decision and Control*, Honolulu, Hawaii, 1990.

[58] S. P. Nicosia, P. Tomei and A. Tornambe, "Approximate Asymptotic Observers for a Class of Nonlinear Systems," *Systems and Control Letters*, Vol. 12, pp. 43–51, 1989.

[59] S. P. Nicosia and P. Tomei, "Observer–based Control Laws for Robotic Manipulators," *International Symposium on Intelligent Robotics*, Bangalore, India, pp. 313–321, 1991.

[60] C. Canudas de Wit and J-J. E. Slotine, "Sliding Observers for Robot Manipulators," *Automatica*, Vol. 27, pp. 859–864, 1991.

[61] J-J. E. Slotine, J. K. Hedrick and E. A. Misawa, "On Sliding Observers for Nonlinear Systems," *ASME J. Dyn. Syst., Meas. and Contr.*, Vol 109, pp. 245–252, 1987.

[62] E. A. Misawa and J. K. Hedrick, "Nonlinear Observers: A State–of–the–Art Survey," *ASME J. of Dyn. Syst., Meas. and Contr.*, Vol. 111, pp. 344–352, 1989.

[63] S. Panzieri and G. Ulivi, "Design and Implementation of a State Observer for a Flexible Robot," *IEEE International Conference on Robotics and Automation*, Atlanta, GA, pp. 204–209, 1993.

[64] D. C. Nemir, A.J. Koivo and R.L. Kashyap, "Pseudo–Links and Self–Tuning Control of a Nonrigid Link Mechanism," *IEEE Transactions on Systems, Man, Cybernetics*, Vol. 18, No. 1, pp. 40–48, Feb. 1988.

[65] R. V. Patel and M. Toda, "Qualitative Measures of Robustness for Multi-variable Systems," *Joint Automatic Control Conference*, San Francisco, CA, No. TP8-A, 1980.

[66] J-J.E. Slotine and W. Li, *Applied Nonlinear Control*, Prentice Hall, Englewood Cliffs, NJ, 1991.

[67] H. K. Khalil, *Nonlinear Systems*, Macmillan, New York, 1992.

[68] J. Descusse and C. Moog, "Decoupling with Dynamic Compensation for Strong Invertibel Affine Nonlinear Systems," *International Journal of Control*, Vol. 42, No. 6, pp. 1387–1398, 1985.

[69]   M. Moallem, R. V. Patel and K. Khorasani, "An Inverse Dynamics Control Strategy for Tip Position Tracking of Flexible Multi–link Manipulators," *Journal of Robotic Systems*, Vol. 14, No. 9, pp. 649–658, Sept. 1997.

[70]   M. Moallem, K. Khorasani and R. V. Patel, "An Integral Manifold Approach for Tip Position Tracking of Flexible Multi–link Manipulators," *IEEE Transactions on Robotics and Automation*, Vol. 13, No. 6, pp. 823-837, Dec. 1997.

[71]   M. Moallem, K. Khorasani and R. V. Patel, "Optimum Structure Design for Flexible–link Manipulators," *IEEE International Conference on Robotics and Automation*, Minneapolis, MN, pp. 798–803, 1996.

[72]   D. N. Godbole and S. S. Sastry, "Approximate Decoupling and Asymptotic tracking for MIMO Systems," *IEEE Transactions on Automatic Control*, Vol. 40, No. 3, pp. 441-450, March 1995.

[73]   D. B. Stewart, D. E. Schmitz and P. K. Khosla, *Chimera 3.2: The Real-Time Operating System for Reconfigurable Sensor-Based Control Systems*. The Robotics Institute, Dept of Electrical and Computer Engineering, Carnegie Mellon University, 1993.

[74]   W. H. Press, B. P. Flannery, S. A. Teukolsky and W. T. Vetterling, *Numerical Recipes in C: The Art of Scientific Computing*. Cambridge University Press, New York, 1995.

[75]   V. I. Utkin, "Variable Structure Systems with Sliding Modes," *IEEE Transactions on Automatic Control*, Vol. AC–22, No. 2, pp. 212–222, 1977.

[76]   R. A. DeCarlo, S. H. Zak and G. P. Mathews, "Variable Structure Control of Nonlinear Multivariable Systems: A Tutorial," *Proceedings of the IEEE*, Vol. 76, No. 3, 1988.

[77]   J. J. Slotine and S. S. Sastry, "Tracking Control of Non–linear Systems Using Sliding Surfaces, with Application to Robot Manipulators," *International Journal of Control*, Vol. 38, No. 2, pp. 465–492, 1983.

[78]   M. Moallem, R. V. Patel and K. Khorasani, "An Observer–based Inverse Dynamics Control Strategy for Flexible Multi–link Manipulators," *35th IEEE Conference on Decision and Control, Kobe, Japan*, 1996.

[79] A. De Luca, P. Lucibello and G. Ulivi, "Inversion Techniques for Trajectory Control of Flexible Robot Arms," *Journal of Robotic Systems*, Vol. 6, No. 4, pp. 325–344, 1989.

[80] J.-J. E. Slotine and J. A. Coetsee, "Adaptive Sliding Controller Synthesis for Non–linear Systems," *International Journal of Control*, Vol. 43, No. 6, pp. 1631–1651, 1986.

[81] P. V. Kokotović, H. K. Khalil, and J. O'Reilly, *Singular Perturbation Methods in Control: Analysis and Design*, Academic Press, 1986.

[82] F-Y. W. Wang and J. L. Russell, "Optimum Shape Construction of Flexible Manipulators with Total Weight Constraint," *IEEE Transactions on Systems, Man, and Cybernetics*, Vol. 25, No. 4, pp. 605–614, 1995.

[83] C. de Boor, *Spline Toolbox for use with MATLAB*, The Math Works, Inc., Natick, MA, 1990.

[84] C. de Boor, *A Practical Guide to Splines*, Applied Mathematical Sciences, Vol. 27, Springer–Verlag, New York, 1978.

[85] A. Grace, *Optimization Toolbox User's Guide*, The Math Works, Inc., Natick, MA, 1990.

[86] W. J. Book and M. Majette, "Controller Design for Flexible, Distributed Parameter Mechanical Arms via State Space and Frequency Domain Techniques," *ASME J. of Dyn. Sys., Meas., and Contr.*, Vol. 105, pp. 245–254, 1983.

[87] J. Hauser and R. M. Murray, "Nonlinear Controllers for Non–integrable Systems: The Acrobot Example," *American Control Conference*, pp. 669–671, 1990.

[88] J. Hauser, S. Sastry and G. Meyer, "Nonlinear Control Design for Slightly Nonminimum–Phase Systems: Application to V/STOL Aircraft," *Automatica*, 28(4), pp. 665–679, 1992.

[89] S. A. Bortoff and M. W. Spong, "Pseudolinearization of the Acrobot using Spline Functions," *31st IEEE Conference on Decision and Control*, Tucson, AZ, pp. 593-598, 1992, pp. 593-598.

[90] K. Hashtrudi–Zaad and K. Khorasani, "An Integral Manifold Approach to Tracking Control for a Class of Non–minimum Phase Linear Systems by Using Output Feedback," *Automatica*, Vol. 32, No. 11, pp. 1533-1552, November 1996.

[91]   M. Moallem, R. V. Patel and K. Khorasani, " Experimental Results for Nonlinear Decoupling Control of Flexible Multi–link Manipulators," *IEEE International Conference on Robotics and Automation,* Albuquerqe, NM, 1997.

[92]   M. Moallem, K. Khorasani and R.V. Patel, "Inverse Dynamics Sliding Control of Flexible Multi–link Manipulators," *American Control Conference,* Albuquerqe, NM, 1997.

[93]   Talebi, H. A., *Neural Network-Based Control of Flexible-link Manipulators,* Ph.D. Dissertation, Department of Electrical and Computer Engineering, Concordia University, Montreal, Canada, 1997.

[94]   M. Moallem, K. Khorasani and R.V. Patel, "Inversion–Based Sliding Control of a Flexible–link Manipulator," *International Journal of Control,* Vol. 71, No. 3, pp. 477-490, 1998.

[95]   W. Gevarter, "Basic Relations for Control of Flexible Vehicles," AIAA Journal, Vol. 8, No. 4, pp. 666-672, 1970.

[96]   O. Khatib, "Reduced Effective Inertia in Macro-Mini Manipulator Systems," *Japan-U.S.A. Symposium on Flexible Automation,* pp. 329-334, 1989.

[97]   W-W Chiang, R. Kraft and R. H. Cannon, "Design and Experimental Demonstration of Rapid, Precise End-Point Control of a Wrist Carried by a Very Flexible Manipulator," *International Journal of Robotics Research,* Vol. 10, No. 1, pp. 30-40, 1991.

[98]   A. R. Fraser, and R. W. Daniel, *Perturbation Techniques for Flexible Manipulators,* Kluwer Academic Publishers, Norwell, MA, 1991.

[99]   M. Moallem, K. Khorasani and R.V. Patel, "Optimum Structure Design for Improving the Dynamics of Flexible–link Manipulators," *International Journal of Robotics and Automation,* Vol. 13, No. 4, pp. 125-131, 1998.

[100] A. Konno, M. Uchiyama, Y. Kito and M. Murakami, "Vibration Controllability of Flexible Manipulators," *IEEE Int. Conf. on Robotics and Automation,* San Diego, CA, 1994.

[101] A. Konno, M. Uchiyama, Y. Kito and M. Murakami, "Configuration-Dependent Vibration Controllability of Flexible-link Manipulators", *International Journal of Robotics Research,* Vol. 16, No. 4, pp 567–576, 1997.

[102] M. Balas, "Feedback Control of Flexible Systems", *IEEE Tran Auto. Control,* 1978.

[102] M. Tarokh, "Measures for Controllability, Observability, and Fixed Modes," *IEEE Transactions on Automatic Control*, Vol. 37, No. 8, pp. 1268–1273, 1992.

[103] G. Kreisselmeier and R. Steinhauser, "Systematic Control Design by Optimizing a Vector Performance Index," *IFAC Symposium on Computer Aided Design of Control Systems*, Zurich, Switzerland, 1979.

[104] D.S. Watkins, *Fundamentals of Matrix Computations,* John Wiley and Sons, New York, 1991.

[105] J.H. Wilkinson, *The Algebraic Eigenvalue Problem,* Oxford University Press, Oxford, 1965.

[106] T. Williams, "Transmission Zero Bounds for Large Space Structures with Applications", *AIAA J. Guidance and Control*, Vol. 12, No. 1, pp. 33–38, 1989.

[107] M.W. Spong, "Underactuated Mechanical Systems,", in *Control Problems in Robotics and Automation*, Lecture Notes in Control and Information Sciences 230 Springer-Verlag, London, UK, 1997.

[108] E.F. Crawley and J. de Luis, "Use of Piezoelectric Actuators as Elements of Intelligent Structures," *AIAA Journal*, 25(10), 1373-1385, 1987.

[109] D. Sun and J.K. Mills, "Study of Piezoelectric Actuators in Control of a Single-link Flexible Manipulator", *IEEE Conf. on Robotics and Autom.*, 849-854, Detroit, MI, 1999.

[110] B. Patnaik, G.R. Heppler, D. Wang, "Stability Analysis of a Piezoelectric Vibration Controller for an Euler-Bernoulli Beam", *American Control Conf.*, San Francisco, CA, 1992.

Index

Acrobot, 21
Admissible functions
    clamped-mass, 7
Aircraft
    PVTOL, 22
Attractive manifold, 99

B-Spline
    coefficients, 115
Beam equation
    Euler-Bernoulli, 6
Boundary layer, 99

Colocated
    sensors and actuators, 10
Computed torque, 11
Control
    composite, 33
    corrective, 33
    fast, 36
    input-shaping, 13
    intelligent, 14
    inverse dynamics, 69

variable structure, 97
Control input
    dynamic, 17
    static, 17
Control law
    implementation, 52, 83
Controllability, 75, 106
    measure, 109
    modal, 109
Controller
    causal, 32

Damping
    ratio, 69
    Rayleigh, 67
    structural, 59, 109
Decoupling
    input-output, 28
Deflection
    tip, 59
    variables, 60
Design

mechanical, 3, 14, 15
    structure, 28
Diffeomorphism, 61
    local, 19
Dynamics
    rigid, 10
    aircraft, 24
    error, 39
    internal, 19, 108
    nonlinear, 2
    unobservable, 14
    zero, 3, 25

Eigenvalue sensitivity, 107

Feedback linearization, 14
Finite-element, 8
Flexibility
    structural, 1
Flexible-link
    model, 9
Flexural rates
    observation strategy, 28
Frobenius norm, 106, 110

Input-output
    description, 75
Integral manifolds, 13, 27, 31,
    35

Lagrangian, 5
Linearization
    input–output, 76
    input-output, 60
    pseudo, 74
Lyapunov analysis, 64

Manifold
    approximate, 36
    condition, 35
    second order, 36, 37

slow, 35
Manipulation
    exact, 2
    high–speed, 2
Manipulators
    flexible, 24
    flexible-link, 5
    rigid-link, 5
    flexible joint, 33
    Macro-Micro, 25
Mass matrix
    ill-conditioned, 86
Mechanical system
    under–actuated, 21
Minimum-phase
    behavior, 62
    local, 62
Mode shapes
    clamped-free, 7
Model
    infinite dimensional, 4
Modeling
    dynamic, 5
Modes
    rigid-body, 108

Non-minimum phase
    characteristic, 11, 24
Nonlinear system
    affine, 64
Nonlinearity
    saturation, 88

Observability, 106
    measure, 109
Observer
    full-order, 76
    reduced-order, 78, 88
    sliding, 79
Operating system

real-time, 49
Optimization, 16
    algorithm, 114
    minmax, 111, 116
    multi-objective, 105
Optimization index
    vector, 117
Output feedback, 37
    static, 20
    dynamic, 31, 41
Output redefinition, 12, 28, 58
Outputs
    fast, 38
    slow, 38

Passivity, 8
Pinned-free
    eigenfunctions, 8
Pole-zero
    cancellation, 113
    locations, 113
    relationship, 112

Reconfigurable subsystems, 84
Reflected-tip, 8
Regulation, 12, 17
Relative degree, 18
Residual set, 64
Rigidity, 2
Robustness, 42

Sensitivity matrix, 109
Shape
    structural, 3, 16
Shape design
    structural, 106
Shape functions
    modal, 5
Singular perturbations, 14, 31
Sliding surface, 98

Stability
    internal, 2
Stability analysis
    Lyapunov, 42
States
    internal, 25
Subsystem
    corrected fast, 37, 41
    corrected slow, 40
    exact fast, 37
    slow, 36
Switching surface, 99
System
    minimum-phase, 19
    nonminimum-phase, 19
        examples, 20

Tracking, 17
    problem, 17
    tip–position, 17
Tracking errors, 45
    tip position, 65

Undamped pendulum, 23

Vibrations
    tip-position, 13

Zero dynamics, 39, 62
    asymptotically stable, 38
    unstable, 22
Zeros
    transmission, 16, 20, 107,
        114

Lecture Notes in Control and Information Sciences

Edited by M. Thoma

1997–2000 Published Titles:

Vol. 222: Morse, A.S.
Control Using Logic-Based Switching
288 pp. 1997 [3-540-76097-0]

Vol. 223: Khatib, O.; Salisbury, J.K.
Experimental Robotics IV: The 4th
International Symposium, Stanford, California,
June 30 - July 2, 1995
596 pp. 1997 [3-540-76133-0]

Vol. 224: Magni, J.-F.; Bennani, S.;
Terlouw, J. (Eds)
Robust Flight Control: A Design Challenge
664 pp. 1997 [3-540-76151-9]

Vol. 225: Poznyak, A.S.; Najim, K.
Learning Automata and Stochastic
Optimization
219 pp. 1997 [3-540-76154-3]

Vol. 226: Cooperman, G.; Michler, G.;
Vinck, H. (Eds)
Workshop on High Performance Computing
and Gigabit Local Area Networks
248 pp. 1997 [3-540-76169-1]

Vol. 227: Tarbouriech, S.; Garcia, G. (Eds)
Control of Uncertain Systems with Bounded
Inputs
203 pp. 1997 [3-540-76183-7]

Vol. 228: Dugard, L.; Verriest, E.I. (Eds)
Stability and Control of Time-delay Systems
344 pp. 1998 [3-540-76193-4]

Vol. 229: Laumond, J.-P. (Ed.)
Robot Motion Planning and Control
360 pp. 1998 [3-540-76219-1]

Vol. 230: Siciliano, B.; Valavanis, K.P. (Eds)
Control Problems in Robotics and Automation
328 pp. 1998 [3-540-76220-5]

Vol. 231: Emel'yanov, S.V.; Burovoi, I.A.;
Levada, F.Yu.
Control of Indefinite Nonlinear Dynamic
Systems
196 pp. 1998 [3-540-76245-0]

Vol. 232: Casals, A.; de Almeida, A.T. (Eds)
Experimental Robotics V: The Fifth
International Symposium Barcelona,
Catalonia, June 15-18, 1997
190 pp. 1998 [3-540-76218-3]

Vol. 233: Chiacchio, P.; Chiaverini, S. (Eds)
Complex Robotic Systems
189 pp. 1998 [3-540-76265-5]

Vol. 234: Arena, P.; Fortuna, L.; Muscato, G.;
Xibilia, M.G.
Neural Networks in Multidimensional
Domains: Fundamentals and New Trends in
Modelling and Control
179 pp. 1998 [1-85233-006-6]

Vol. 235: Chen, B.M.
H∞ Control and Its Applications
361 pp. 1998 [1-85233-026-0]

Vol. 236: de Almeida, A.T.; Khatib, O. (Eds)
Autonomous Robotic Systems
283 pp. 1998 [1-85233-036-8]

Vol. 237: Kreigman, D.J.; Hagar, G.D.;
Morse, A.S. (Eds)
The Confluence of Vision and Control
304 pp. 1998 [1-85233-025-2]

Vol. 238: Elia, N. ; Dahleh, M.A.
Computational Methods for Controller Design
200 pp. 1998 [1-85233-075-9]

Vol. 239: Wang, Q.G.; Lee, T.H.; Tan, K.K.
Finite Spectrum Assignment for Time-Delay
Systems
200 pp. 1998 [1-85233-065-1]

Vol. 240: Lin, Z.
Low Gain Feedback
376 pp. 1999 [1-85233-081-3]

Vol. 241: Yamamoto, Y.; Hara S.
Learning, Control and Hybrid Systems
472 pp. 1999 [1-85233-076-7]

Vol. 242: Conte, G.; Moog, C.H.; Perdon A.M.
Nonlinear Control Systems
192 pp. 1999 [1-85233-151-8]

Vol. 243: Tzafestas, S.G.; Schmidt, G. (Eds)
Progress in Systems and Robot Analysis and Control Design
624 pp. 1999 [1-85233-123-2]

Vol. 244: Nijmeijer, H.; Fossen, T.I. (Eds)
New Directions in Nonlinear Observer Design
552pp: 1999 [1-85233-134-8]

Vol. 245: Garulli, A.; Tesi, A.; Vicino, A. (Eds)
Robustness in Identification and Control
448pp: 1999 [1-85233-179-8]

Vol. 246: Aeyels, D.;
Lamnabhi-Lagarrigue, F.; van der Schaft, A. (Eds)
Stability and Stabilization of Nonlinear Systems
408pp: 1999 [1-85233-638-2]

Vol. 247: Young, K.D.; Özgüner, Ü. (Eds)
Variable Structure Systems, Sliding Mode and Nonlinear Control
400pp: 1999 [1-85233-197-6]

Vol. 248: Chen, Y.; Wen C.
Iterative Learning Control
216pp: 1999 [1-85233-190-9]

Vol. 249: Cooperman, G.; Jessen, E.; Michler, G. (Eds)
Workshop on Wide Area Networks and High Performance Computing
352pp: 1999 [1-85233-642-0]

Vol. 250: Corke, P. ; Trevelyan, J. (Eds)
Experimental Robotics VI
552pp: 2000 [1-85233-210-7]

Vol. 251: van der Schaft, A. ; Schumacher, J.
An Introduction to Hybrid Dynamical Systems
192pp: 2000 [1-85233-233-6]

Vol. 252: Salapaka, M.V.; Dahleh, M.
Multiple Objective Control Synthesis
192pp. 2000 [1-85233-256-5]

Vol. 253: Elzer, P.F.; Kluwe, R.H.; Boussoffara, B.
Human Error and System Design and Management
240pp. 2000 [1-85233-234-4]

Vol. 254: Hammer, B.
Learning with Recurrent Neural Networks
160pp. 2000 [1-85233-343-X]

Vol. 255: Leonessa, A.; Haddad, W.H.; Chellaboina V.
Hierarchical Nonlinear Switching Control Design with Applications to Propulsion Systems
152pp. 2000 [1-85233-335-9]

Vol. 256: Zerz, E.
Topics in Multidimensional Linear Systems Theory
176pp. 2000 [1-85233-336-7]